THE CLOCK
OF THE LONG NOW

OTHER WORKS
BY STEWART BRAND:

Whole Earth Catalog
Two Cybernetic Frontiers
The Media Lab: Inventing the Future at MIT
How Buildings Learn: What Happens After They're Built

THE CLOCK
OF
THE LONG NOW

Time and Responsibility

STEWART BRAND

BASIC
BOOKS

A Member of the Perseus Books Group

Published by Basic Books,
A Member of the Perseus Books Group

Designed by Rachel Hegarty

A CIP catalog record for this book is available from the
Library of Congress.
ISBN 0-465-04512-X (cloth); ISBN 0-465-00780-5 (pbk.)

CONTENTS

CONTENTS

THE CLOCK
OF THE LONG NOW

NOTIONAL CLOCK

Time and Responsibility. What a prime subject for vapid truisms and gaseous generalities adding up to the world's most boring sermon. To spare us both, let me tie this discussion to a specific device, specific responsibility mechanisms, and specific problems and cases. The main problems might be stated, How do we make long-term thinking automatic and common instead of difficult and rare? How do we make the taking of long-term responsibility inevitable?

The device is a Clock, very big and very slow. For the purposes of this book it is strictly notional, a Clock of the mind, an instrument for thinking about time in a different way. As it happens, such a Clock is in fact being built. The builders are finding that the very idea of the Clock—why to build it, how to build it—forces their thinking in interesting directions; among other things, toward long-term responsibility. Since it works for them, please consider yourself one of the Clock's builders. It won't take long to catch up. Here's a project summary from late 1998, complete with preamble:

> Civilization is revving itself into a pathologically short attention span. The trend might be coming from the acceleration of technology, the short-horizon perspective of market-driven economics, the next-election perspective of democracies, or the distractions of personal multitasking. All are on the increase. Some sort of balancing corrective to the short-sightedness is needed—some mechanism or myth that encourages the long view and the taking of long-term responsibility, where "the long term" is measured at least in centuries.
>
> What we propose is both a mechanism and a myth. It began with an observation and idea by computer designer Daniel Hillis, who wrote in 1993:
>
>> When I was a child, people used to talk about what would happen by the year 2000. Now, thirty years

later, they still talk about what will happen by the year 2000. The future has been shrinking by one year per year for my entire life.

I think it is time for us to start a long-term project that gets people thinking past the mental barrier of the Millennium. I would like to propose a large (think Stonehenge) mechanical clock, powered by seasonal temperature changes. It ticks once a year, bongs once a century, and the cuckoo comes out every millennium.

Such a clock, if sufficiently impressive and well engineered, would embody deep time for people. It would be charismatic to visit, interesting to think about, and famous enough to become iconic in the public discourse. Ideally, it would do for thinking about time what the photographs of Earth from space have done for thinking about the environment. Such icons reframe the way people think.

Hillis, who developed the "massive parallel" architecture of the current generation of supercomputers, has devised the mechanical design of the Clock and is now building the prototype. Its works consist of an ingenious binary digital-mechanical system that has precision equal to one day in twenty thousand years, and it self-corrects by *phase locking* to the noon sun. For the way the eventual Clock is experienced (its size, housing, etc.), we expect to keep proliferating design ideas for a while. The prototype Clock, only eight feet tall, is shaping up beautifully in Monel alloy, Invar alloy, tungsten carbide, metallic glass, and synthetic sapphire.

The Clock project became the Clock/Library with the realization of the need for content to go along with the long-term context provided by the Clock—a "library of the deep future, for the deep future." The Clock/Library could take care of kinds of information deemed especially useful over long periods of time, such as minding extremely long-term scientific studies, or accumulating a Responsibility Record of policy decisions with long-term consequences.

To deliver mythic depth, the Clock/Library needs to be a remarkable facility at a remarkable location. High deserts are attractive for their broad horizons and high-preservation climate, if the specific location is not too remote for easy access worldwide. City sites offer high visibility but are harder to protect over centuries. Our strategy is to develop a city Clock/Library first—for visibility—and then a desert Clock/Library—for longevity. Before both comes a World Wide Web site (at www.longnow.org).

In addition to its mythic core facility the Clock/Library as a cultural tool may need to be widely dispersed—on the Net, in publications and distributed services, and at various locations. The point, after all, is to explore whatever may be helpful for thinking, understanding, and acting responsibly over long periods of time. Manifestations of the overall project could range from fortune cookies to theme parks. For now, we will build an astonishing Clock and a unique Library and see what develops from there.

Who is "we"? The Long Now Foundation was established in 1996 to foster long-term responsibility. The founding board is Daniel Hillis (co-chair), Stewart Brand (co-chair), Kevin Kelly, Douglas Carlston, Peter Schwartz, Brian Eno, Paul Saffo, Mitchell Kapor, and Esther Dyson. Hillis created Thinking Machines Inc. and its supercomputer, the Connection Machine, and is now a Fellow at Disney. Brand began the *Whole Earth Catalog* and co-founded Global Business Network. Kelly is executive editor of *Wired* magazine and author of *Out of Control*. Carlston co-founded Broderbund Software. Schwartz is chairman of Global Business Network and author of *The Art of the Long View*. Eno is a musician, music producer, and artist. Saffo is spokesman for Institute for the Future. Kapor founded Lotus and co-founded the Electronic Frontier Foundation. Dyson created and runs *Release 1.0*, the leading computer industry newsletter.

The present version of the Clock/Library scheme has grown from three years' online conversation among the board members. Brian Eno proposed "the long now" as what we are aiming to promote. Peter Schwartz suggested 10,000 years as the appropriate time envelope for the project: 10,000 years ago was the end of the Ice Age and beginning of agriculture and civilization; we should develop an equal perspective into the future. Douglas Carlston noted that the institution to maintain this project will be as much of a design challenge to last one hundred centuries as the Clock or the Library.

As of autumn 1998 The Long Now Foundation has an executive director (Alexander Rose), two staffers, an office at the San Francisco Presidio, and nonprofit status. The foundation is developing its funding, the web site, the working prototype of the Clock, small conferences on subjects such as "Managing Digital Continuity" (that one was at the Getty Center in Los Angeles, February 1998), conception of the Library and its initial services, and locations to build.

What's your advice? Where should the Clocks be located? How should they be experienced? How should the Library work? What kind of institution could maintain them for 10,000 years?

Do wade into the conversation, if you like. There is an appendix at the back of the book spelling out how to contact the Clock/Library project directly. In this book are a number of voices beside mine, here in part to encourage your thoughts and invite your voice.

Whether or not a grand version of the Clock eventually happens, the world continues to happen, and it happens to be in a new scale of trouble these days. Nobody can save the world, but any of us can help set in motion a self-saving world—if we are willing to engage the processes of centuries, because that is where the real power is.

The Clock provides a framework for this book. The book in turn aims, as a by-product, to provide a framework for the actual Clock. Though the chapters are written in logical sequence, they do not attempt a sequential and conclusive argument. They are a mosaic; many are quite short, some differ in situational voice (a fictional speech, a genuine paper). Each of the chapters is a different kind of probe, a separate essay. The overall intent is exploratory rather than convergent. These are early days. Thinking in ten-thousand-year terms is new to us. We have a long way to go to comprehend even the size of the subject of very long-term responsibility.

KAIROS AND CHRONOS

The number of human beings now alive is around six billion. The estimated number of humans who have ever been alive is about one hundred billion. What is the number of humans who *will* be alive? We owe the past humans our existence, our skills, and our not-bad world. What do we owe the future humans? Existence, skills, and a not-bad world. Maybe even a better world.

Over just the next one hundred years, extrapolating from United Nations population projections, 12.6 billion new humans will be having lives—twice as many as those of us alive now. Their century, 2000 to 2100 C.E.,* is the world of our children and grandchildren. We are used to thinking of the 12.6 billion strictly as devouring mouths, but they are mainly our heirs, people we care about and for whom we save and invest.

Since the soon-to-be outnumber the living; since the living have greater impact on the unborn than ever before thanks to depletion of natural systems, atmospheric disruption, toxic residue, burgeoning technology, global markets, genetic engineering, and sheer population numbers; since our scientific and historic understandings now comfortably examine processes embracing eons; and now that our plan-ahead horizon has shrunk to five years or less—it would seem that a grave disconnect is in progress. Our ever-hastier decisions and actions do not respond to our long-term understanding, or to the gravity of responsibility we bear.

"The greatest good for the greatest number" means the *longest* good, because the majority of people affected is always yet to come. We can do little good for our dead, but immeasurable good—or harm—to our unborn. Accepting responsibility for the health of the whole planet, we are gradually realizing, also means responsibility

---

*C.E., meaning "of the common era," is rightly considered preferable to A.D. (*anno Domini*, "in the Year of the Lord"), which has become objectionably narrow and Christian in a fast-globalizing world. In this preferred usage B.C.E. ("before the common era") replaces B.C. ("before Christ").

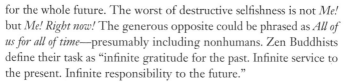
for the whole future. The worst of destructive selfishness is not *Me!* but *Me! Right now!* The generous opposite could be phrased as *All of us for all of time*—presumably including nonhumans. Zen Buddhists define their task as "infinite gratitude for the past. Infinite service to the present. Infinite responsibility to the future."

Responsibility, these years, means mastering long lead times, long lag times, and the hidden effects of cumulative change. In the domain of atmosphere and climate the delay between cause and effect can be thirty years. One climatologist notes, "We are the first generation that influences global climate, and the last generation to escape the consequences." Ecological communities are tremendously resilient, but the current human-caused extinction rate (one hundred to one thousand times normal) steadily erodes their resilience until the communities suddenly crash, irreparably and without warning. It would be convenient if we could shrug off the degrading of ecosystems, but our own health depends on them for water, food, air, and temperate weather. In a *Science* article titled "Human Domination of Earth's Ecosystems," the authors observe, "We are changing Earth more rapidly than we are understanding it." Their counsel is to back off a little: "Ecosystems and the species they support may cope more effectively with the changes we impose, if those changes are slow."

The deer frozen in the headlights, the driver frozen at the wheel with no time to brake or swerve—both are doomed by speed and bad luck. Luck you cannot do much about; speed you can. Overdriving the headlights—that is, counting on no surprises out there in the darkness—is folly on any road. Braking time must match awareness time.

Patricia Fortini Brown, in *Venice and Antiquity*, notes that the ancient Greeks distinguished two kinds of time, "*kairos* (opportunity or the propitious moment) and *chronos* (eternal or ongoing time). While the first . . . offers hope, the second extends a warning." *Kairos* is the time of cleverness, *chronos* the time of wisdom.

Our dead and our unborn reside in the realm of *chronos*, murmuring warnings to us presumably, if we would ever look up from our opportunistic, *kairo*tic seizures of the day. This must be the Golden Age of *Kairos* we live in. Or the Mercury Age of *Kairos*—fluid as quicksilver, shimmering . . .

Poisonous.

Thrilling.

MOORE'S
WALL

I t didn't start as a law, it started as a prediction. In retrospect it turned out to be the most accurate and consequential prediction in the history of technology, and it exposed the structure of technological hyperacceleration in the late twentieth century. In the April 19, 1965, issue of the technical journal *Electronics*, it was one of a series of papers called "The Experts Look Ahead." The author, head of semiconductor research and development at Fairchild Camera and Instrument Corporation, was an electrical engineer with a background in physical chemistry. The title of his three-and-a-half-page article was "Cramming More Components onto Integrated Circuits." Mostly a technical discussion of the advantages of integrated circuits (computer chips), it did loft a few astounding pronouncements: "The future of integrated electronics is the future of electronics itself. . . . Integrated circuits will lead to such wonders as home computers—or at least terminals connected to a central computer—automatic controls for automobiles, and personal portable communications equipment. The electronic wristwatch needs only a display to be feasible today."

The author was Gordon E. Moore, later cofounder of Intel Corporation, the world's leading computer chip manufacturer. What came to be known as Moore's Law was a small graph and explanation buried in his *Electronics* paper. From 1965 Moore looked back to the beginnings of integrated circuits in 1959 and noted that the number of components (transistors) that could be fit on a chip had doubled every year for six years. He predicted that the trend would continue for another ten years, permitting an astonishing sixty-five thousand components on a chip by 1975. (The actual numbers by 1975 were around twelve thousand, so the formula was later adjusted to predict doubling every eighteen months.)

History veered—not only as a result of the power of the new technology of computation, communication, and intelligence but also owing to the self-accelerating rate of its arrival described by Moore's Law. Dense computer chips were used to make still denser

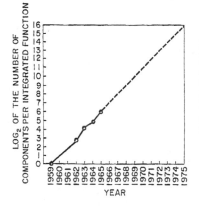

FIGURE 3.1 Gordon Moore's 1965 graph of the growing number of components on microchips used what is called a *semilog scale*. The horizontal scale of time is linear (arithmetic), while the vertical scale of number of components is logarithmic ($\log_2$), also called *exponential*. It counts the number of doublings, going from one component per chip in 1959 to 64 components per chip in 1965 (six doublings), and extrapolating—figuring one doubling every year—to 2,048 components per chip in 1970 (eleven doublings) and 65,536 components per chip in 1975 (sixteen doublings). The actual number in 1975 turned out to be 12,000 components per chip (a little over ten doublings), and so Moore's Law was later amended to say that the number of components per chip doubled every eighteen months.

That number—doubling every eighteen months—set the pace of technological change for fifty years.

The effect on humans, who do not live exponentially, has been more like what is displayed on the right; *both* the vertical and horizontal scales are linear. Moore's Law becomes Moore's Wall.

1959

1999

computer chips, ad infinitum. Doubling the number of components on a chip every eighteen months kept doubling computer power and halving expense. Computers would just keep getting smaller, faster, cheaper, and smarter, and not at a steady rate but explosively. The explosion burst past 1975, continued through 1985, 1995, and it shows every sign of constant acceleration to at least 2015: thirty-seven doublings, about a 137 billionfold increase of power in fifty-six years. There is no precedent in the history of technology for the sustained self-feeding growth of computer capability.

The pace of Moore's Law has become the pacesetter for human events. According to a rule of thumb among engineers, any tenfold quantitative change *is* a qualitative change, a fundamentally new situation rather than a simple extrapolation. Moore's Law brings such tenfold structural changes every three years or so, thus three revolutions every decade, for five decades straight.

One revolution after another swept through the computer and communications industries. Personal computers took off in the mid 1980s, displacing mainframes and minicomputers. There were "Moore's Laws" of doubling capacity in digital storage, in communications bandwidth, and in the ubiquity of microprocessors in everything from dolls and doorknobs to hearing aids. New machines became obsolete every three years. The proliferation of personal computers and the digitizing of communications via the Internet set off what came to be called Metcalfe's Law, named after Xerox engineer Bob Metcalfe. It states that the power of a network grows as the square of the number of users (people or devices) on the net. Hence the explosion of the World Wide Web in the mid 1990s, when the Net's total content was doubling every one hundred days.

Related technologies such as biotech also took off. By 1997 Monsanto Corporation claimed that a computer-accelerated "Monsanto's Law" was operating at a similar pace: "The ability to identify and use genetic information is doubling every twelve to twenty-four months. This exponential growth in biological knowledge is transforming agriculture, nutrition and health care in the emerging life sciences industry." Monsanto chose not to mention human germline engineering—that is, designing disease-resistant, higher-yield people.

Velocity itself became the dominating characteristic of the world's quicksilver economy. "We are moving from a world in

The Systematic Value of Compatibly Communicating Devices Grows As the Square of Their Number:

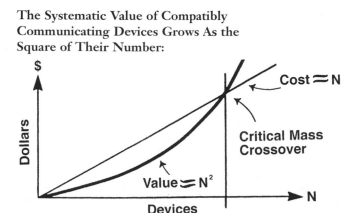

FIGURE 3.2 Bob Metcalfe's original diagram shows how the value of a net increases as the square of the number of nodes (or people) on the net. It was drawn in 1973 at Xerox PARC (a corporate research center in California) while he was working on a local-area network system called Ethernet. The diagram shows that the costs of a network system are linear, whereas the growing value of the net is exponential, and that therefore the value will surpass the costs at the "critical mass crossover" and thereafter ascend to glory.

The formula $V = n^2$ is more precisely $V = n(n-1)$. Imagine a net of 10 people. Each of them has 9 other people as resources. Total value (V) is 90. Double the number of people and you get approximately quadruple the value: with 20 people, each has 19 others as resources; total value 380. Ten times the number of people, 100 times the value: with 100 people, each has 99 resources; total value 9,900. A thousand times the people, a million times the value. What was the value of the Internet in 1998, when it had some 50 million people on it?

Metcalfe's Law explains why 50 million people *had* to get on the Internet in just a few years. The aggregate value of other users was so great that they could not afford to miss the boat.

which the big eat the small," remarked Klaus Schwab, head of the World Economic Forum, "to a world in which the fast eat the slow."

With technological acceleration driving economic acceleration, politics and culture can only struggle to keep up. Armed conflict changes. There's no time for the boredom that generates military or political adventuring, such as in the 1960s; instead the relentless pace of events inspires conservative wars or revolutions that try to slow things down. Now that we have progress so rapid that it can be observed from year to year, no one calls it progress. People call it change, and rather than yearn for it, they brace themselves against its force.

Technology is treated as something that pushes us around rather than something we create. It's a bother, it's a boon, it's a discipline; it's a given. "What people mean by the word *technology*," says computer designer Alan Kay, "is anything invented since they were born." Computer designer Danny Hillis counters, "What people mean by the word *technology* is the stuff that doesn't really work yet." Technology is both the problem and its own solution. No wonder it obsesses us.

The gathering acceleration of history was noted in the 1790s by Thomas Malthus and in 1909 by Henry Adams, who wrote,

> The world did not double or treble its movement between 1800 and 1900, but, measured by any standard . . . the tension and vibration and volume and so-called progression of society were fully a thousand times greater in 1900 than in 1800—the force had doubled ten times over, and the speed, when measured by electrical standards as in telegraphy, approached infinity, and had annihilated both space and time.

What would Adams say about the year 2000? We speak in his terms about the events of decades now, not centuries. One advantage of that, perhaps, is that the acceleration of history now makes us all historians. Observing structural-level change directly and personally, we strive to understand it so we can prepare for whatever comes next.

Old people and young people live in completely different time zones. Watching old movies, we are struck by how slow-paced they seem. On Wall Street the standard buy-sell cycle has gone from two years to three months. In business, management consultants mock the past: "Five-year plan??? For managers who get it, it's more like a five-week schedule inside a five-month plan inside of a fifteen-month intuition." (The stating of Moore's Law, after all, was originally as a business objective.)

"If Moore's Law is true," queries a media developer, "over time is time more or less valuable?" In other words, is compressed time dearer or more disposable? The price per minute is higher, but is the sustainable value? Does intense progress make everything better, or just more temporary?

Moore's Law, in its own grotesque way, is a constant, something that planners now routinely take into account. But we're used to living arithmetically (1, 2, 3, 4, 5, 6 . . . 40), not exponentially (1, 2, 4, 8, 16, 32 . . . one trillion). Later doublings in an exponential sequence, we come to realize, are absolutely ferocious. The changes no longer feel quantitative or qualitative but cataclysmic; each new doubling is a new world. Ad man Regis McKenna calls it "continuous discontinuous change." Life becomes perpetual transition with no resting point in sight.

Danny Hillis explains what motivated him to build a linear Clock in an exponential era: "Some people say that they feel the future is slipping away from them, but to me the future is a big tractor-trailer slamming on its brakes in front of me just as I was pulling into its slipstream. I am about to crash into it. It feels like something big is about to happen—all those graphs showing the yearly growth of population, atmospheric concentration of carbon dioxide, number of Net addresses, and megabytes per dollar. They all soar up to an asymptote just beyond the turn of the century: the Singularity."

THE SINGULARITY

The metaphor of the Singularity comes from astrophysics. What makes it so compelling to futurists and trend watchers? Like any effective metaphor, it hides distracting elements (such as how good something is supposed to be for you) and reveals hidden properties that are essential. The Singularity metaphor answers the question, What happens if our technology just keeps accelerating?

Beyond a certain critical mass an expiring giant star collapses not only to a superdense neutron star but to something whose mass and density is so great that its intense gravitational force makes the escape velocity of anything from the object greater than the speed of light. The collapsed star becomes what is called a *black hole*. The region where light and everything else disappears from our universe into the black hole is termed the *event horizon*. The beyond-dense anomaly in the center of the black hole is called a *singularity*. "At this singularity," writes the Cambridge mathematician Stephen Hawking, "the laws of science and our ability to predict the future would break down."

The man who applied this metaphor to human events is the science fiction writer and mathematician Vernor Vinge. His 1991 novel *Across Realtime* joins three stories he wrote in the mid 1980s around a central mystery: What happened to everybody? While the characters in the stories were temporarily isolated out of time in devices called *bobbles*, civilization and the rest of humanity disappeared from Earth. Reconstructing events leading up to the disappearance, the characters realized that, at the time, technology advance was radically self-accelerating. Innovations that formerly had taken years were being made in months and then days. Then the record stopped. Vinge's characters called the event *the Singularity*—"a place where extrapolation breaks down and new models must be applied. And those new models are beyond our intelligence." In the metaphor radical progress is not progress but the end of the world as we know it. In a 1985 afterword to the original stories Vinge predicted that the Singularity would happen in reality, in the lifetime of his readers.

A good many people, including Clock designer Danny Hillis, have adopted Vinge's term as a shorthand way of referring to impending technology acceleration and convergence. They all note that the future becomes drastically unpredictable beyond the Singularity. Among some enthusiasts there is even a consensus date for what they call the *techno-rapture*—2035 C.E., give or take a few years.

Opinions vary as to what will be the Singularity's leading mechanism. Proponents of nanotechnology (molecular engineering) are sure that the turning point will be "the assembler breakthrough"— that is, when ultratiny, ultrafast nanomachines capable of self-replication are devised. Others expect that the convergence of computer technology, biotechnology, and nanotechnology, each accelerating the other, will fuse into a new order of life. Vinge himself sees the tipping point as the moment when machine intelligence, or machine-enhanced intelligence, surpasses normal human intelligence and takes over its own further progress. Another possibility is some emergent property of the all-embracing Internet, which Vinge proposes might "suddenly awaken."

Any such occurrence would indeed transform our world. Whether or not it will actually occur, the mere prospect of a technological Singularity changes behavior. People already refer to the near future in months instead of years, and to the distant future in years instead of decades or centuries. What may happen decades from now—beyond the imagined event horizon—is treated as not only unknown but unknowable. Under such conditions speed becomes glorified. Haste switches from a vice to a virtue; behavior that once might have been called reckless and irresponsible becomes swift and decisive action.

One reason the metaphor resonates is that it offers insight about the distortion we all feel from the pace of events these years. As one falls into a black hole, the fierce gravitational field of the singularity pulls the traveler into a long thin shape, like taffy. The more the accelerating future pulls at us the more parts of us resist; the result is a kind of dismemberment.

Society itself could be dismembered, as some people ride the breaking wave of ever newer technology over the event horizon into invisibility while others lag behind, feeling the powerful gravitational force of still-accelerating technology yet no longer able to

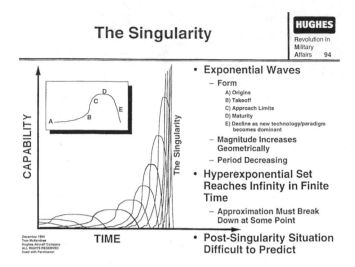

FIGURE 4.1 A version of the Singularity focusing on multiple successive technology paradigms. Each new technology is increasingly powerful, and each gives way to a new paradigm in a shorter time. Instead of capability growing exponentially over uniform time (as with Moore's Law), the ever-shortening time periods makes the aggregate power curve "hyperexponential."

This presentation slide was prepared in 1994 by Tom McKendree at Hughes Aircraft. It was used for discussions about military contracting in an era of what was called the Revolution in Military Affairs: the anticipated effects of high-tech acceleration on the tools and conduct of war.

see it. Thus the world would be comprehensible only to those near the leading edges of technology.

The Singularity is a frightening prospect for humanity. I assume that we will somehow dodge it or finesse it in reality, and one way to do that is to warn about it early and begin to build in correctives.

I used to skydive—until I had a parachute fail—so I can vouch for the following account. I asked Rosalind Picard, a professor at MIT's Media Lab, if there was any time dimension in emotions. She wrote back:

> I remember when skydiving—actually "flying" during freefall—that a couple minutes in the sky spanned a mental day of adventure. Skydivers who mount cameras on their heads to capture the events in the air have to shoot at twice the speed at which they plan to play the video. This is necessary for things to look "normal." The brain, emotionally high on adrenaline, runs much faster during freefall than when viewing the video afterward. In fact, the video viewed "normalized" in this way tends to recreate that adrenaline rush in the viewer.

The word *freefall* is a pretty good descriptor for our times. It conveys the thrill of danger, the speeded-up rush, the glorious freedom, and the fall. The secondary reason I quit skydiving was one day seeing it from outside. When you watch someone jump from your plane they are out on the step grinning fiercely into the propwash, they push off, and they become quickly tiny and distant. When it is you jumping, however, you push the plane away, it becomes tiny and distant, and your jumpsuit rattles violently in the 120-mile-per-hour wind. From within the experience what you see is exciting but not frightening. Then I watched someone jump from an adjoining plane. From the side view I saw reality: The person just *plummeted*.

So besides having a hell of a good time skydiving, I learned three visceral-level lessons. One, never count totally on equipment. My parachute malfunction was statisically inevitable. Two, always have a backup. I remember remarking to myself as I pulled my reserve chute, "I hope this one works. I don't have any more parachutes." I would have lived just eight seconds more if it hadn't opened. Three, look for an outside frame of reference. This may explain my interest in the Clock/Library project—the Clock is the outside frame of reference; the Library is the backup.

A late 1990s headline in the *Christian Science Monitor* read, "SPREAD OF TECHNOLOGY GIVES RISE TO A CULTURE OF IMMEDIACY." A British credit card company called Access had as its advertising slogan, "Access takes the waiting out of wanting." A book from the period began, "Imagine a world in which time seems to vanish and space becomes completely malleable. Where the gap between need or desire and fulfillment collapses to zero." This was in *Real Time*, by Regis McKenna. "Never mind if it has become a cliché to accuse American business and Wall Street of too much short-term thinking," he added. "The positive side of this trait is a hair-trigger responsiveness as measured against the standards of European or Japanese companies. . . . In real time, the best is the enemy of the good." In other words, if you take time to perfect your product, you'll be too late to market.

To be in business is to hustle—hustle the customer and hustle faster than the competition. Electronically accelerated market economies have swept the world for good reasons. They are grassroots driven (by customers and entrepreneurs), swiftly adaptive, and highly rewarding. But among the things they reward, as McKenna points out, is short-sightedness.

One of the least reported, least reflected upon trends of the late twentieth century has been the rise of gambling. Growing at a rate of about 20 percent a year through the 1990s in the United States, the amount annually spent on legal gambling passed $700 billion in 1998. About 8 percent of that went to the "house"—$56 billion in profits, bigger than the domestic film and music industries combined. Instead of curtailing the game government joined it, actively teaching citizens to bet unthinkingly. States with lotteries went from one in 1964 to thirty-seven in 1997. The number of addicted gamblers increased accordingly, along with the usual crime, broken families, and suicides. The gaming industry has become a powerful political lobby, buying government acquiescence and media silence.

The clangorous flood of coins into the bin of your slot machine is such a rush. The sweet release of cocaine hitting your bloodstream is such a rush. Exciting times. But the price of the routine rush is futureless melodrama, in personal life as in business. There is no outside perspective, no side view, no story larger than "now what?" and self-characterization ("I'm a wild and crazy guy").

After years of working with alcoholics the anthropologist and psychologist Gregory Bateson observed, "If the hangover preceded the binge, drunkenness would be considered a virtue and not a vice." Drunks disbelieve in the inevitability of the hangover and the harm they cause while drunk. Their vice is dismissing the future and surrendering to the rewards of the moment, the binge. Bateson's idea of hangover-first would reward and teach patience. The reality of binge-first annihilates both patience and memory.

If taking thought for the future was essential in steady times, how much more important is it in accelerating times, and how much harder? It becomes both crucial and seemingly impossible. There are so many new varieties of short-term opportunity, and the pace of events buffets our attention with so many surprises, it is as if the old dialogue between opportunistic *kairos* and durational *chronos* has become a monologue, just a shriek of joy into the gale of freefall.

THE
LONG NOW

In the early winter of 1979 Brian Eno was taking a break from pop stardom with what became a five-year interlude in New York City, reconnecting with nonpop innovation in music and art. He recalls,

> I was having a great time, but I kept feeling there was a conceptual poverty of a particular kind in the society I moved within. Exactly what this was became clear to me one day when I was invited over to the very glamorous loft of a celebrity—a $2 million design job located in a rough part of town. We climbed over the bums on the doorstep, having bumped our way down the dreadful streets in a crappy taxi, and walked into this vision of total decadent luxury. During dinner I asked the hostess, "Do you like living here?" "Oh sure," she replied, "this is the loveliest place I've ever lived."
>
> I realized that the "here" she lived in stopped at her front door. This was a very strange thought to me. My "here" includes the neighborhood at least. After that, I noticed that young arty New Yorkers were just as local in their sense of "now." "Now" meant "this week." Everyone had just got there, and was just going somewhere else. No one had any investment in any kind of future except their own, conceived in the narrowest terms.
>
> I wrote in my notebook that December, "More and more I find I want to be living in a Big Here and a Long Now." I guess part of the reason the idea attracted me is that it offered a justification for the type of music I was starting to make at the time—a music which was sort of suspended in an eternal present tense."

(Eno called it "ambient music." His albums, such as *Music for Airports*, drew critical sneers. Fifteen years later a whole genre of pop music called Ambient claimed Eno as its parent.)

In a 1994 discussion of how to think about and name Danny Hillis's millennial Clock, Eno suggested, "How about calling it 'The Clock of the Long Now,' since the idea is to extend our concept of the present in both directions, making the present longer? Civilizations with long nows look after things better. In those places you feel a very strong but flexible structure which is built to absorb shocks and in fact incorporate them."

How long *is* now, usually? In the Clock discussion Esther Dyson suggested, "On the stock exchange it's today, on the Net it's a month, in fashion it's a season, in demographics a decade, in most companies it's the next quarter." The shortest now is performed in a poem by the Polish poet Wislawa Szymborska:

> *When I pronounce the word Future,*
> *the first syllable already belongs to the past.*

For most of us most of the time I think Eno is right: "now" consists of this week, slightly haunted by the ghost of last week. This is the realm of immediate responsibility, one in which we feel we have volition, where the consequences of our actions are obvious and surprises limited. The weekend is a convenient boundary.

The sociologist Elise Boulding diagnosed the problem of our times as "temporal exhaustion": "If one is mentally out of breath all the time from dealing with the present, there is no energy left for imaging the future." In a 1978 paper Boulding proposed a simple solution: expand our idea of the present to two hundred years—a hundred years forward, a hundred years back. A personally experienceable, generations-based period of time, it reaches from grandparents to grandchildren—people to whom we feel responsible. Boulding, a mother of five, wrote that a two-hundred-year present "will not make us prophets or seers, but it will give us an at-homeness with our changing times comparable to that which parents can have with an ever-changing family of children as they move from age to age."

Two hundred years is good; there is emotional comfort and behavioral discipline in it. If what we want is to change mind-set, however, two hundred years is too readily imaginable, too incremental. Frames of mind change by jumps, not by degrees.

Ten thousand years is the size of civilization thus far. In that time a number of civilizations and dozens of empires have risen and fallen or receded, but the overall advance and convergence of civilization on the planet has been steady. It has reached the point where people now talk comfortably about the emergence of global governance, not by conquest but by merging of interests.

The trigger was agriculture. By 8000 B.C.E. the ice had receded from most of the Northern Hemisphere. Starting sometime before 7000 B.C.E. in the Mideast—where, according to Jared Diamond, the greatest number of plants and animals were candidates for domestication—formerly nomadic tribes settled down around their new crops. Villages formed; some grew into cities, and cities are where civilization happens. States and empires became the standard control apparatus for managing regional economics (e.g., markets, taxes) and resources (e.g., irrigation, food storage). They co-opted religion as a behavioral stabilizer and built armies for use as internal police as well as combat with other empires.

Over time the empires and religions told larger and longer stories about themselves, many of them overlapping and referring to one another. Around 1000 B.C.E., thanks to the Jews, history emerged as another form of storytelling. Eventually archaeology was able to revive stories and histories thought dead, such as those of ancient Egypt and the Mayans. Humanity now shares a complex of stories reaching all the way back to the villages; this is our Long Now.

Ten thousand years is not all that long. It is only four hundred generations—counting a new generation every twenty-five years, four to a century, for a hundred centuries. "Ancient" Egypt was relatively late in the game. The pharaohs started two hundred generations ago (3000 B.C.E.) and built their greatest pyramids within seventeen generations, about the same time as Stonehenge (2575 B.C.E.).

Those original farmers ten millennia ago were the first systemic futurists. They mastered the six-month lag between sowing and reaping, and they remembered enough crop experience and matched it with enough astronomy to be able to use the sky as an accurate signal of when to plant. Such tricks confer advantage. Agriculture-based civilizations replaced hunter-gatherers and in time were able to prevail over even the fiercest marauders.

Other long-term frames of reference may be used as well. Geologically, the last ten thousand years is the Holocene—the thin slice in the Quaternary period of the Cenozoic era at the top of the stratigraphic epoch charts. In astronomical terms civilization is microscopic. It is best measured in comets, such as Halley's, whose seventy-five- to seventy-nine-year return has been documented for twenty-two centuries. The name "Halley's" is only constant since 1759 C.E. The comet named "Hale-Bopp" in 1997 C.E., when it put on a dazzling show, was previously seen in 2214 B.C.E.; we do not know what it was called then. When it returns in 4377 C.E., will anyone mention the name "Hale-Bopp"? Returning comets will let us know whether civilization is developing more continuity of knowledge or less.

Might humanity pay consistent attention through one complete precession of the equinoxes, as the Earth's axis pirouettes around a point in the sky near the Pole Star? This 25,784-year cycle is known as the Great Year. How about keeping track through one rotation of our galaxy—220 million years? The Earth has existed for nearly twenty-five of those galactic rotations, life on Earth for nineteen rotations. Humans may well eventually affect the periodicity of ice ages—we have been frozen by one every one hundred thousand years for a million years and are now enjoying an "interglacial" period—but it seems unlikely that we will have much influence on the rotation of our galaxy or anyone else's, nor will we tally their spin. The human time frame is narrower than that of life, of the planet, and of galaxies.

Eno's Long Now places us where we belong, neither at the end of history nor at the beginning, but in the thick of it. We are not the culmination of history, and we are not start-over revolutionaries; we are in the middle of civilization's story.

The trick is learning how to treat the last ten thousand years as if it were last week, and the next ten thousand as if it were next week. Such tricks confer advantage.

THE ORDER OF CIVILIZATION

In civilizations with long nows, says Brian Eno, "you feel a very strong but flexible structure . . . built to absorb shocks and in fact incorporate them." One can imagine how such a process evolves: All civilizations suffer shocks, yet only those that absorb the shocks survive. This still does not explain the mechanism, however.

In recent years a few scientists (such as R. V. O'Neill and C. S. Holling) have been probing a similar issue in ecological systems: How do they manage change, and how do they absorb and incorporate shocks? The answer appears to lie in the relationship between components in a system that have different change rates and different scales of size. Instead of breaking under stress like something brittle these systems yield as if they were malleable. Some parts respond quickly to the shock, allowing slower parts to ignore the shock and maintain their steady duties of system continuity. The combination of fast and slow components makes the system resilient, along with the way the differently paced parts affect each other. Fast learns, slow remembers. Fast proposes, slow disposes. Fast is discontinuous, slow is continuous. Fast and small instructs slow and big by accrued innovation and occasional revolution. Slow and big controls small and fast by constraint and constancy. Fast gets all our attention, slow has all the power. All durable dynamic systems have this sort of structure; it is what makes them adaptable and robust.

Consider, for example, a coniferous forest. The hierarchy in scale of pine needle, tree crown, patch, stand, whole forest, and biome is also a time hierarchy. The needle changes within a year, the tree crown over several years, the patch over many decades, the stand over a couple of centuries, the forest over a thousand years, and the biome over ten thousand years. The range of what the needle may do is constrained by the tree crown, which is constrained by the patch and stand, which are controlled by the forest, which is controlled by the biome. Nevertheless, innovation percolates throughout the system via evolutionary competition among lin-cages of individual trees dealing with the stresses of crowding, parasites, predation, and weather. Occasionally, large shocks such as fire or disease or human predation can suddenly upset the whole system, sometimes all the way to the biome level.

The mathematician and physicist Freeman Dyson makes a related observation about human society:

> The destiny of our species is shaped by the imperatives of survival on six distinct time scales. To survive means to compete successfully on all six time scales. But the unit of survival is different at each of the six time scales. On a time scale of years, the unit is the individual. On a time scale of decades, the unit is the family. On a time scale of centuries, the unit is the tribe or nation. On a time scale of millennia, the unit is the culture. On a time scale of tens of millennia, the unit is the species. On a time scale of eons, the unit is the whole web of life on our planet. Every human being is the product of adaptation to the demands of all six time scales. That is why conflicting loyalties are deep in our nature. In order to survive, we have needed to be loyal to ourselves, to our families, to our tribes, to our cultures, to our species, to our planet. If our psychological impulses are complicated, it is because they were shaped by complicated and conflicting demands.

In terms of quantity there are a great many pine needles and a great many humans, many forests and nations, only a few biomes and cultures, and but one planet. The hierarchy also underlies much of causation and explanation. On any subject, ask a four-year-old's annoying sequence of *Why?* five times and you get to deep structure. "Why are you married, Mommy?" "That's how you make a family." "Why make a family?" "It's the only way people have found to civilize children." "Why civilize children?" "If we didn't, the world would be nothing but nasty gangs." "Why?" "Because gangs can't make farms and cities and universities." "Why?" "Because they don't care about anything larger than themselves."

Considered operationally rather than in terms of loyalty, I propose six significant levels of pace and size in the working structure of a robust and adaptable civilization. From fast to slow the levels are:

- Fashion/art
- Commerce
- Infrastructure
- Governance
- Culture
- Nature

In a healthy society each level is allowed to operate at its own pace, safely sustained by the slower levels below and kept invigorated by the livelier levels above. "Every form of civilization is a wise equilibrium between firm substructure and soaring liberty," wrote the historian Eugen Rosenstock-Huessy. Each layer must respect the different pace of the others. If commerce, for example, is allowed by governance and culture to push nature at a commercial pace, all-supporting natural forests, fisheries, and aquifers will be lost. If governance is changed suddenly instead of gradually, you get the catastrophic French and Russian revolutions. In the Soviet Union government tried to ignore the constraints of culture and nature while forcing a Five-Year-Plan infrastructure pace on commerce and art. Thus cutting itself off from both support and innovation, the USSR was doomed.

We can examine the array layer by layer, working down from fast and attention-getting to slow and powerful. Note that as people get older, their interests tend to migrate to the slower parts of the continuum. Culture is invisible to adolescents but a matter of great concern to elders. Adolescents are obsessed by fashion, elders bored by it.

The job of fashion and art is to be froth: quick, irrelevant, engaging, self-preoccupied, and cruel. Try this! No, no, try *this!* Culture is cut free to experiment as creatively and irresponsibly as society can bear. From all this variety comes driving energy for commerce (e.g., the annual change in automobile models) and the occasional good idea or practice that sifts down to improve deeper levels, such as governance becoming responsive to opinion polls, or culture gradually accepting multiculturalism as structure instead of grist for entertainment.

If commerce is completely unfettered and unsupported by watchful governance and culture, it easily becomes crime, as in some nations and republics after the fall of communism. Likewise, com-

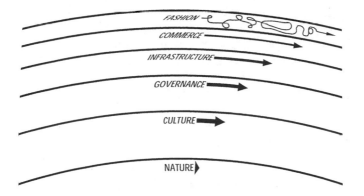

FIGURE 7.1 The order of civilization. The fast layers innovate; the slow layers stabilize. The whole combines learning with continuity.

merce may instruct but must not control the levels below it, because commerce alone is too shortsighted. One of the stresses of our time is the way commerce is being accelerated by global markets and the digital and network revolutions. The proper role of commerce is to both exploit and absorb these shocks, passing some of the velocity and wealth on to the development of new infrastructure, at the same time respecting the deeper rhythms of governance and culture.

Infrastructure, essential as it is, cannot be justified in strictly commercial terms. The payback period for such things as transportation and communication systems is too long for standard investment, so you get government-guaranteed instruments such as bonds or government-guaranteed monopolies. Governance and culture must be willing to take on the huge costs and prolonged disruptions of constructing sewer systems, roads, and communication systems, all the while bearing in mind the health of even slower "natural" infrastructure, such as water, climate, and so on.

Education is intellectual infrastructure; so is science. Very high yield, but delayed payback. Hasty societies that cannot span these delays will lose out over time to societies that can. On the other hand, cultures too hidebound to allow education to advance at infrastructural pace also lose out.

In the realm of governance the most interesting trend in current times besides the worldwide proliferation of democracy and the rule of law is the rise of what is coming to be called the "social sector." The public sector is government, the private sector is business, and the social sector is nongovernmental, nonprofit do-good organizations. Supported by philanthropy and the toil of volunteers, they range from church charities, local land trusts, and disease support groups to global players like the International Red Cross and World Wildlife Fund. What such organizations have in common is that they serve the larger, slower good.

The social sector acts on culture-level concerns in the domain of governance. One example is the sudden mid-twentieth-century dominance of historic preservation of buildings, pushed by such organizations as the National Trust for Historic Preservation in America and English Heritage and the National Trust in Britain. Through them culture declared that it was okay to change clothing at fashion pace but not buildings; okay to change tenants at commercial pace but not buildings; okay to change transportation at infrastructure pace but not neighborhoods. "If some parts of our society are going to speed up," these organizations seemed to say, "then other parts are going to have to slow way down, just to keep balance." Even New York City, the most demolition-driven metropolis in America, began to preserve its downtown.

Culture is where the Long Now operates. Culture's vast slow-motion dance keeps century and millennium time. Slower than political and economic history, it moves at the pace of language and religion. Culture is the work of whole peoples. In Asia you surrender to culture when you leave the city and hike back into the mountains, traveling back in time into remote village culture, where change is century-paced. In Europe you can see it in terminology, where the names of months (governance) have varied radically since 1500 but the names of signs of the Zodiac (culture) remain unchanged for millennia. Europe's most intractable wars are religious wars.

As for nature, its vast power, inexorable and implacable, continues to surprise us. The world's first empire, the Akkadian, in the Tigris-Euphrates valley, lasted only a hundred years, from 2300 B.C.E. to 2200 B.C.E. It was wiped out by a drought that went on for three hundred years. Europe's first empire, the Minoan civilization, fell to earthquakes and a volcanic eruption in the fifteenth century B.C.E. When we disturb nature at its own scale—as with our "extinction engine" and greenhouse gases of recent times—we risk triggering apocalyptic forces. Like it or not, we now have to comprehend and engage the still Longer Now of nature.

The division of powers among the layers of civilization allows us to relax about a few of our worries. We should not deplore rapidly changing technology and business while government controls, cultural mores, and so-called wisdom change slowly; that's their job. Also, we should not fear destabilizing positive-feedback loops (such as the Singularity) crashing the whole system. Such disruption usually can be isolated and absorbed. The total effect of the pace layers is that they provide many-leveled corrective, stabilizing negative feedback throughout the system. It is precisely in the apparent contradictions of pace that civilization finds its surest health.

OLD-TIME RELIGION

Of all cultural practices, religion is the greatest sustainer and most durable of institutions. Invoking a higher power, religion motivates people to step out of their immediate self-interest and serve a higher good. Through rituals at birth, marriage, and death, it helps people think in terms of whole lifetimes and generations.

As you read this book, every day, every few hours—at matins, lauds, prime, terce, sext, and compline—the Catholic divine office is said by monks in Benedictine, Trappist, and Carmelite monasteries. "They are not following time, but sustaining it," writes French philosopher Michel Serres. "Their shoulders and their voices, from biblical verses to orisons, bear each minute into the next throughout duration, which is fragile and would break without them." Mechanical clocks were first invented for monasteries in the late thirteenth century and only later ordered the life of towns. The miniature clock strapped to your wrist is the direct descendent of European monastic practice. It was the monks who taught us to keep time.

Yet something is missing in religious time—deliberately and proudly missing.

To appreciate the timescale of the following Jainist account (India, sixth century B.C.E.) it helps to know that an "ocean of years" is one hundred million times one hundred million palyas. Each palya is a period of countless years.

> This age, known as Very Beautiful, Very Beautiful, lasted 400 trillion oceans of years, and gave way to that known as Very Beautiful, which—as the name suggests—was exactly half as fortunate as the former. The wish-fulfilling trees, the earth, and the waters were only half as bountiful as before. Men and women were only four miles tall, had only 128 ribs, lived for only two periods of countless years, and passed to the world of the gods when their twins were only 64 days

old. This period lasted 300 trillion oceans of years, declining gradually but inevitably to the stage called Sorrowfully Very Beautiful, when joy became mixed with grief.

There is awe, dazzle, delight, humility, and escape in such numbers. But as a way to think about time, sociologist Elise Boulding dryly observes,

> The breathing in and out of the universe by Brahma every four thousand million years is not an image of the future calculated to motivate record-keeping, planning, and action, nor has it done so in India. The lack of interest in history in Indian culture has long been complained of by historians, who can't locate records because no one bothered to keep them. A supposed Indian lack of interest in the future is also sometimes complained of by development planners.

Religious time is time out. Time out from personal striving or suffering, time out from the chaos of history. In the sacred place set apart, in the sacred ritual changeless and timeless, in the sacred communion with a higher order, we step out of ordinary time and thereby make life meaningful, or at least bearable.

From the perspective of the sacred, history is just one damned thing after another. At best it consists almost entirely of bad news. At worst it is sin. In any case it is illusion. The only good news, the only redemption, the only reality abides in transcendent timelessness, in the eternal. Eternity is the opposite of a long time. In his 1949 study of the religious beliefs and practices of archaic societies, *The Myth of the Eternal Return*, Mircea Eliade was struck by "their revolt against concrete, historical time, their nostalgia for a periodical return to the mythical time of the beginning of things, to the 'Great Time.'"

Theocratic civilizations also scanted history. In ancient Egypt one pharaoh would casually claim the historic victories of another two hundred years previous. The pharaoh's job was to maintain eternal order, not to author or suffer unique events. The canon of Egyptian statuary (e.g., the ratios of wrist to elbow to armpit, etc.) was absolutely rigid for 2,200 years—twenty-three dynasties, eighty-eight generations of artists. Such a canon was the opposite of a medium for historic portraiture.

The Jews consecrated their own history, including their historic encounter with Egypt, and thus introduced the idea of history to the world. Still they sought release from history, placing the escape in the future. All the "peoples of the Book" adopted varieties of this approach. To caricature each of their stances: Judaism says, "The Messiah is going to come, and that's the end of history"; Christianity says, "The Messiah is going to come back, and that's the end of history"; Islam says, "The Messiah came; history is irrelevant."

One Sunday morning at a Long Now Foundation board retreat on Clock/Library design, Kevin Kelly, a devout Christian, spoke up: "I go to church, but why am I here and not in church? It's because I feel that the Christian church denies the future. From year one, they have been waiting for the second coming. I think we need a story that includes the future."

CLOCK/
LIBRARY

"8. Provide a description of what a visitor to the clock, library, etc. would see and experience."

Alexander Rose groaned. How could he articulate to the U.S. Internal Revenue Service a fantasy that was changing on a weekly basis? As executive director of The Long Now Foundation, he had been getting from the IRS nothing but bureaucratic questions and a year of delays in granting the foundation its crucial 501 (c) 3 public education nonprofit status. Forcing the issue, one of Long Now's potential donors had said he would not write a check until the IRS approval came through.

Rose reviewed his previous correspondence with the IRS—all of his careful, understated, bureaucratically hedged wording. Frustrated, he gave up and just blurted out pure dream this time, incorporating some of the elements that Long Now board members were discussing that month, adding embellishments of his own. You want a detailed answer? Here's a detailed answer:

8. We will describe what we currently hope to be the experience of the full-size Clock/Library complex. You exit your vehicle in a parking area at the base of a mountain somewhere in the high desert of the Southwest United States. Looking up, you see a flight of shallow steps, each step carved from a layer of rock representing approximately 10,000 years of geologic time. After climbing one hundred of these steps, or one million years into the future, you are somewhat awed and belittled by the greatness of geologic time.

You arrive at a flat knoll where you see a cave ahead. Through the opening of the cave you see some large but slow movement. You proceed and gradually make out a giant pendulum swinging back and forth deep within the cave. Once you reach the center you realize that you are actually within the clock mechanism itself and you are aware of the pendulum beating out its 10-second period. You proceed up a spiral staircase that will take you through the relatively low ceiling and up into the first layer of clock mechanics. On this layer you see the fastest of the mechanical calculation devices, which ticks once per day. As you go up flight after flight you see each progressive mechanism with its relatively slower tick, the last being the precession of the equinoxes, a 25,784-year cycle. The next few layers are the abstraction layers that adjust solar time to actual time and the delay for the pendulum-impulsing mechanism.

When you reach the top of the stairs you are in a huge room several stories tall. It is dimly lit from a slot cut through the living rock of the mountain on the southern face. You make out two giant helices, one descending either wall, each being rotated by a falling weight that must weigh several tons. Then you are surprised by an immediate brightness in the room. It is coming from the sun that has just become directly in line with the slit on the wall. It is reflecting off a hemispherical mirror lighting up the whole room and heating up a sphere in the center of a great dial. The heating of this sphere actuates a synchronization mechanism which automatically adjusts the time of the clock to local noon. You are able to make out the dial around this sphere, now showing you the year in the cryptic method of keeping time when this clock was built. It reads the year 11,567. You then look at the rings in from this to find images you recognize of the Sun and Moon in their current phases, as well as a diagram of the current night sky. From these you are able to work backward the actual time to your newer and more familiar time scale. But you are struck that the people of this ancient time had the foresight to think this far into their future and create this place.

At this point you wander through the rest of the facility to find a library and people accessing and preserving the data stored there. Akin to the truly ancient library of Alexandria, there is a constant forward migration of the data to increasingly better and denser methods of storage. In the main vault you find the original 1,000 books stored at the impossibly large scale of 100 nanometer pixels. These were the first 1,000 books stored in the Clock/Library chosen by its founders. Although not necessarily relevant to your time, what they began helped to teach people the value of knowledge over long periods of time. Without it humanity might have obsolesced itself out of existence without being able to look over the ancient records of the sea and air and find trends that are only apparent over centuries or millennia.

IRS approval of Long Now's nonprofit status came back by return mail.

The lesson? Don't hold back. Clock/Library is an extravagant project. The more over the top it is the better it works. If it wants to compete for iconic power with the mushroom cloud and the photos of Earth from space, it has to match their vaulting ambition. (Let's split the atom! Let's leave Earth!) The great time-spanning precedent is the pyramids of Giza. Their massive monumentality defined when "history" began. They are a stabilizer, a frame of reference, for any culture that cares to care about them. The pyramids also demonstrate the power of folly. You can't argue with them because they're not rational.

The ambition and folly of the Clock/Library is to reframe human endeavor, and to do so not with a thesis but with a thing. All this thing can do is give permission to think long term. If it succeeds in that, the rest may follow.

The Clock's instigator and designer, Danny Hillis, thinks of the project in terms of his three children, who want to know what their story is and where they might fit into it. "In some sense, we've run out of our story, which was the story of taking power over nature. It's not that we've finished that, but we've gotten ahead of ourselves, and we don't know what the next story is after that." The Clock is a way of bridging between stories, embodying respect for the full span of the old story and confidence in the gradual emergence of a new story. It is a transition-managing device. In a world of hurry the Clock is a patience machine. Says Hillis, "If you're going to do something that's meant to be interesting for ten millennia, it almost has to have been interesting for ten millennia. Clocks and other methods of measuring time have interested people for a very long time."

Dan Wolf, one of the people with whom Hillis worked at Disney, commented, "A traditional clock depicts time in the context of

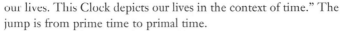

our lives. This Clock depicts our lives in the context of time." The jump is from prime time to primal time.

The main characteristic of the Clock is its linearity. It treats one year absolutely like another, oblivious of Moore's Law accelerations, national fates, wars, dark ages, or climate changes. In its company there is nothing special about now. While we discount on a sliding scale both the future and the past, the Clock does neither. Far future and near future are the same; distant past and recent past have equal value. In times of turbulence the Clock emanates calm. In calm times it reminds us that no equilibrium is stable for long.

Clock/Library aims for the mythic depth to become, as Brian Eno puts it, "one of those system-level ideas which sets in motion all sorts of behavior without ever having to be referred to directly again. This is what dominant myths do: they make some sorts of behavior ring with recognition and familiarity and value and a sense of goodness, and thus lay deep templates for social cohesion about what would otherwise be very hard-to-discuss topics."

For the Clock to be such a point of reference it must deliver an experience. After an encounter with the Clock a visitor should be able to declare with feeling, "Whew. *Time!* And me in it." It is not so much a conversion experience as a deep pause, like coming upon the Grand Canyon by surprise, where you simply want to sit and watch for a while and let your life adjust to two million years visible in one glance. Yet the Grand Canyon, like the night sky, is crushing in scale; there is no way to engage it personally. The Clock's time frame of four hundred generations is human scale—it should invite you to engage it personally, by doing something, leaving some mark.

In the high mountain passes of the world generations of travelers build impressive rock cairns, each stone a token of thanks for safe passage, a humorously small addition to the great total of the cairn. If visitors can make some kind of investment in the process of the Clock, there is a greater chance of their investing in its portent, which might include feelings such as "There is time to sort our problems out," or "I'm part of something long and lively," or "There is always an alternative to my present urgency—and it's not a vacation, it's acknowledging deeper responsibility."

Your ideas are welcome on ways in which the Clock might invite participation; here are some of the early ones. Hillis: "My current thinking is that the Clock should be powered by people coming to visit it. If it is forgotten, it stops." Eno: "I've always loved prayer wheels. You see these things in the walls of Muslim cities. People walking by spin them. We need some system of recording that kind of human contribution. Perhaps people have to turn little wheels to keep certain essential parts of the Clock going. If they fail to turn them, atrophy begins." The Clock inspires responsibility by inviting some.

Another Eno idea revolves around a sound environment that may encourage visitors to contribute silence:

> Quiet music stops people talking. It makes people aware that they may be intruding on someone else's experience if they talk loudly. It slows them down, makes them realize they're having an experience which exists in time, has duration, and that therefore they might want to stop shuffling around and sit still for a bit to wait for the experience to unfold.
>
> Years ago I did some recordings where I just put mikes at the end of long cardboard tubes and then put the other ends of the tubes out the windows—such that each tube pointed off in a different direction. Everything that happened outside then made the tubes "ring" at their resonant frequencies. This set up a constant ringing chord, the intensity and timbre of individual notes within the chord varying according to what was going on outside each particular window. I wonder whether a bigger version of this could make a constant (and constantly changing) soft humming sound in the Clock chamber. The attraction of course is that this is a piece of music that requires no energy and no maintenance. The longer the tube, the deeper the fundamental resonance (i.e., the basic note). Even quite short tubes will create subsonic frequencies, so what you'd be hearing with a long tube (for instance, fifty feet) would be a harmonic series above the fundamental, and from that (by dint of the psychological effect called "fundamental tracking") your brain would deduce that there was a very low fundamental note.

This is the kind of subtle yet simple design idea the Clock should collect and integrate. The concept, as well as the experience, of "fundamental tracking" is particularly appealing, since it is in a way what the Clock is trying to do: make us sense the deep, inaudible note of centuries by giving us the audible harmonics.

What would you like to see in and around such a Clock? The musician Peter Gabriel suggests that visitors make life-mask bricks of themselves that get added to the growing structure. Someone's thought about wind chimes turned into the more impressive idea of earthquake chimes. The naturalist Peter Warshall proposed growing giant crystals in the Clock; a large amethyst would take about two thousand years. Eno would like to paint the floor of highly trafficked areas with layers of different-colored paint so that wear shows up quickly and beautifully. I want to peddle a century-watch that has only a second hand (to show that it's running) and a hand that tells what century you're in.

While the Clock's tangibility inspires specific conjuring, the idea of the Library is more nebulous, slower to take shape, slower to generate value, though over time it might accrue great value. Hillis sees the Library as the Clock's evolutionary companion:

> To me the Clock and the Library capture two different as-
> pects of time. The Clock is Newtonian time, physical
> time—reversible, regular, steady. The Library is informa-
> tion time. It points in a direction, it grows, it's unpre-
> dictable. One of the fundamental mysteries of the universe
> to me is how one kind of time can be made out of the
> other—how information time can be built on top of physi-
> cal time. At some level, this is the most interesting question
> in physics and in religion. It's the question of where the uni-
> verse is heading.

The two kinds of time might gain savor in each other's pres-
ence. The Clock dramatizes the scope of historic time past and to come but offers no content. The Library is all content, especially past content with future significance. One use of it might be "applied history," a concept offensive to historians because it is by def-
inition nonobjective. People act out of historical understanding anyway; perhaps it could be done better. Some Library services might deal with the future directly. One such service came to Long Now board member Doug Carlston in a dream: messages to the fu-

ture. Clock/Library could provide, for a fee, time-mail service across generations forward.

There are already many wonderful libraries. It would take centuries of imaginative curating and uses for a Millennial Library to acquire unique value. The value could lie in providing civilizations with a wisdom line: slow, robust, apparently inefficient. In this respect it might be like a species' genotype, which contains much more hidden diversity (in recessive genes, mutations, etc.) than what is expressed in the current bodily phenotypes. By its very inefficiency the genotype preserves tremendous adaptivity in the species.

Dark ages come to everyone's mind when thinking about very long-term libraries—at least everyone in the West, because Europe had one. After the fall of Rome formal learning disappeared for half a millennium. Only in the rural Benedictine monasteries were intellectual discourse and education maintained. So it went for five centuries, until suddenly in the mid twelfth century the lead was taken over by the new universities in reviving cities such as Paris, Bologna, and Oxford. Urban universities have kept the intellectual lead ever since (though the Net may be challenging that).

Future dark ages are always possible. The technological Singularity could generate one—what if whatever we transition to doesn't work out? Contemplating the need for bridging gaps in civilization, the dreamers of a millennial Clock/Library assumed from the beginning that it would be built in a remote desert site. Forests, shorelines, and cities suffer too rapid a physical turnover to be good candidates. The recollection of what happened to Europe's monasteries after 1200—they became stagnant and irrelevant—suggests that Clock/Library should not lose touch with urban creativity and ferment. So the overall scheme became bimodal: There would be an urban Clock/Library for zest and ongoing relevance, and a desert Clock/Library for permanence and perspective; they would, in a sense, keep each other honest. (Reprise from the pace-layering Chapter 7: "Fast learns, slow remembers. Fast proposes, slow disposes.")

The strategy so far is to build the Clock quickly, the Library incrementally. Start with a city Clock/Library, then on the basis of that experience assemble the big desert version. The city one would be new for a while. The one in the desert might be built so gradually that it is never new.

All of which still leaves the question, Who tends the Clock and the Library for the ongoing millennia? The clearest model of how to run a charismatic site is the central Shinto complex in Japan known as the Ise Shrine. While Stonehenge has become a pathetic ruin, no one having the faintest idea how it was actually used, and the Giza Pyramids only having survived looting and the extinction of their religion by sheer physical mass, the Ise Shrine is entirely fresh and beautiful, its religion the living spiritual framework of the nation.

The historian Daniel Boorstin reports that the Inner Shrine at Ise was first built in 4 B.C.E., the Outer Shrine in 478 C.E. Every twenty years for well over a thousand years the all-wood shrine has been totally reconstructed—a perfect replica built next to the previous building. The sixty-first rebuilding took place in 1993. Ise is the world's greatest monument to continuity—an unbroken lineage of structure, records, and tradition on a humid, earthquake-prone, volcanic island. Its ancient rites are alive and meaningful. Materials dismantled from the replaced building are recycled to other shrines throughout Japan. "The cakes and sake for the renewal ceremonies," writes Boorstin, "are made from rice ceremonially transplanted in the same seven-acre rice paddies that have been used for two thousand years."

Continuity and perpetual renewal go together. The way to maintain a living monument, Ise proves, is to provide it with a living institution. The Long Now Foundation might become or inspire that kind of institution for the Clock/Library. It should not let itself become a religion, though. That would lead too easily to fanaticism, it would be almost certain to become sectarian, competing with existing religions, and it would invite the kind of future-dodging common to religions.

A helpful motto for the Clock/Library tenders, therefore, would be: "We don't do eternity." What might other guidelines be for such an organization that intends to survive and be valuable for a very long time? Figuring out the answer is likely to be a long conversation indeed, which you are invited to join. For now, The Long Now Foundation has come up with these candidate guidelines:

1. Serve the long view (and the long viewer).
2. Foster responsibility.

3. Reward patience.
4. Mind mythic depth.
5. Ally with competition.
6. Take no sides.
7. Leverage longevity.

Big Ben, the world's largest accurate clock, is a prime example of "sublime" technology: an engineering artifact that inspires public awe and fetishistic fascination. Other sublime works are the Eiffel Tower, Hoover Dam, Golden Gate Bridge, and Saturn 5 rocket that took men to the Moon. The tower of Big Ben is majestic in scale: 314 feet high. The massive machinery of the clock and belfry exemplify Victorian technology. It is visibly (and aurally!) charismatic and readily comprehensible. The bigness of Big Ben is its essence.

One winter day in the mid 1990s I took a tour of Big Ben with Brian Eno. The first part of the experience is climbing the 340 spiral steps; you wind yourself to the top. We found the belfry a high and windy place with stunning views down on the waters of the Thames and the roofs and turrets of Parliament. The monument is named for its biggest bell, Big Ben. He dominates the middle of the belfry—fourteen tons, his note E natural, to toll England's hours. His shape is unusual: 9 feet wide and 7.5 feet high, made that way to fit up the too-narrow air shaft misdesigned by Parliament's architect, Charles Barry, who fought publicly for decades with the clock's designer, C. B. Denison.

Suspended above Big Ben we saw the quarter-hour chiming bells: G sharp (1 ton), F sharp (1.25 tons), E natural (1.75 tons), and B natural (3.5 tons). Most of the chiming clocks in the world play their now-familiar "Westminster" melody, taken from Handel's *Messiah*. It has words:

> *All through this Hour,*
> *Lord be my Guide,*
> *That by Thy Power,*
> *No foot should slide.*

> *Dong ding dang gong; dong dang ding dong;*
> *ding dong dang gong; gong dang ding dong.*

As the hour approaches, the guide warns you to keep your head away from any hard surface, because the shock of the first explosive chime makes many people jerk their head and hurt themselves. But if you watch the F bell you can see its hammer pull back just before smiting the bronze. Such a fine, measured clamor! In the pause between the chimes and the tolling of the hour, Eno leaned toward the hammer (448 pounds) of Big Ben. We saw it hoist, then *WHONNGGG!!!* Eleven big ones.

You can stand it bare-eared, even revel in it. It shakes you physically, like the roar from a space-rocket launch. You are at the core sound of Britain and its abiding myth. Big Ben live strikes the hour on BBC worldwide—its first hour strike exact to the second. This sound achieved its abiding stature when it was broadcast throughout the duration of World War II to a global audience, pealing out freedom and steadfastness for millions.

Eno wrote after the tour,

> What interests me is the fact of this clock being so closely identified with our self-image. Big Ben is a sort of throbbing heart for British culture—calm, assured, implacable, accurate (and thus "just"), enduring, and *big*. This list represents just about all the things Britain used to think it was, and would still like to think it is. We are aware of the measured symbolism of its chimes beaming out *"Fairness, fairness"* across the planet.

Accuracy exact to the second was thought to be impossible in a turret (tower) clock in the early 1800s, when the "gifted amateur" horologist Denison designed Big Ben's clockwork. It would have to drive eight huge hands on four faces out in the wind and ice without any pressure feeding back to the mechanism. To make the energy flow only one way he devised *Grimthorpe's double three-legged gravity escapement* (Denison was later Lord Grimthorpe). The best innovation of its kind in centuries, it became the standard escapement for pendulum clocks, including grandfather clocks. No patents. Denison never did work for money.

It is indeed an impressive mechanism. After the customary stroll inside the translucent clock faces designed by the great neo-Gothicist Augustus Pugin, Eno and I went down a flight to view the works; Big Ben's sublimity lives in their superbly crafted details. Three gear trains and weight drums are arrayed in order on a fifteen-foot cast-iron platform four feet wide. Left to right: the hour-striking mechanism; the *going* (clock) mechanism; and the quarter-hour chime mechanism. The cable drums for the bell mechanisms dominate both ends, because they must convey so much energy to the bell hammers from the huge hanging weights: 1 ton for the hour strike, 1.25 tons for the quarter-hour chimes. The weight for the going train weighs only 560 pounds. All three are wound all the way back up the tower every three days—originally by hand, now by electric motor.

The pendulum hangs from a piece of 1/64-inch-thick spring steel, which has to flex millions of times. It has only been replaced once since Big Ben was installed in 1859. Thirteen feet long from the suspension point to its center of gravity, the pendulum has a stately two-second beat:

TICK.

TOCK.

With every tick the tip of each fourteen-foot minute hand on the four clock faces advances a visible inch. In a year, one hundred miles.

Eno and I dutifully viewed the famous pennies used to adjust the clock, a stack of which is visible on the boss of the pendulum a few feet below the suspension point. Each one-ounce penny speeds the clock by 0.4-second in 24 hours (by raising the pendulum's center of gravity a trace and thus shortening its period a microtrace). The pendulum, made of a zinc rod inside an iron tube, compensates for any changes in temperature and is protected from air currents by a cast-iron, enclosed pendulum pit. The role of the pennies is to correct for barometric changes—higher pressure, denser air: simply add a penny.

At the escapement energy "escapes" from the weight through the gear train to the clock hands. The rotation of the escapement (eighteen inches in diameter) is the most visible motion on the clock—until the quarter hour chimes. Then, what a commotion! A huge overhead flyfan (air brake) thrashes around, gears grind, and a large wheel rotates cams against levers, which pull the cables that raise and drop the hammers on the thundering bells above. Eno was particularly fascinated by the elegance of the fist-sized cams arrayed like music-box pins on the wheel: sculpture that makes music.

What keeps the 560-pound pendulum going? With each swing the weight-driven going train imparts a one-ounce nudge to the pendulum. Everything—the pendulum, the near-perfect measure, the gear trains, the huge clock hands, the great bell hammers—is driven by an astronomical force: gravity. Plus another astronomical force—starlight from the Sun—that, processed by photosynthesis through oil and grain, raises the weights and nourishes the intelligence to design and care for such an instrument.

The best book about Big Ben was written in the mid 1980s by the resident engineer, John Darwin. He remarks in *The Triumphs of Big Ben* that the repairs he oversaw were intended to give the clock two centuries of reliable service before the next major overhaul. How different a future-oriented mechanism is from a mere monument. How much more alive.

THE WORLD'S SLOWEST COMPUTER

If you were going to design a clock to keep good time for 10,000 years, where would you begin? Danny Hillis started with certain design principles (that might apply to other things beside clocks, as does his whole approach of breaking down a problem into components and examining an exhaustive range of options). Invention comes at the end of this process, not at the beginning. The principles:

- Longevity: display correct time for ten millennia.
- Maintainability: with Bronze-Age technology, if need be.
- Transparency: obvious operational principles.
- Evolvability: improvable over time.
- Scalability: the same design should work from tabletop to monument size.

These principles led to a set of design strategies. Achieving longevity suggests: go slow; avoid sliding friction such as gears; avoid impacts such as ticking; stay clean; stay dry; expect bad weather, earthquakes, and human interaction; don't tempt thieves or vandals.

The principles of maintainability and transparency propose the use of familiar materials; make it easy to build spare parts; allow inspection; rehearse motions (so seldom-moving parts don't freeze up); expect restarts; and include the manual or be the manual—that is, build it so that everything is intuitively obvious.

For evolvability and scalability make all parts of similar size; provide a simple interface/readout; separate the functions—power, timing, calculation, and display.

Hillis analyzed each of these four functions in terms of what had been used in clocks before and what might be used now. For the power source water flow and wind are too damaging. Tidal power, plate tectonic, geothermal, chemical, and stored potential energy all scale badly. Atomic and solar electric power are difficult to maintain. Only temperature change (such as a large bimetallic lever bent by the difference between day and night temperatures) and human winding survived the analysis. *Human winding* is the best candidate because it fosters responsibility (one purpose of the Clock, after all) and invites people's involvement.

There are even more options for the timing mechanism. Too inaccurate: pendulum, water flow, solid material flow, spring and mass, wear or corrosion, ball roll, diffusion, audio oscillator, pressure chamber cycle, or inertial governor. Too unreliable: daily temperature cycle, seasonal temperature cycle, solar alignment (problematic clouds), or stellar alignment (clouds). Too difficult to measure: tidal forces, Earth's rotating inertial frame, or tectonic motion. Too high-tech and therefore difficult to maintain over a long time frame: piezoelectric oscillator (as in quartz watches), atomic oscillator.

The unsatisfactory conclusion of this analysis led Hillis to one of his Clock innovations. He would use an unreliable but accurate timer (*solar alignment*) to adjust an inaccurate but reliable timer (*pendulum*), creating a *phase-locked loop*. A Huygens (swinging) pendulum or torsional (rotating) pendulum would keep the Clock close to accurate, and then a pulse of focused sunlight at exact solar noon would adjust the Clock precisely on any day there was sun. In the event of prolonged cloudiness from volcanic eruptions, nuclear winters, or large meteor impacts, the Clock's pendulum could keep close-enough time for a few years until the Sun came out again.

A second major innovation emerged from analyzing the calculation options. Electronic calculation (as in digital clocks and watches) would be hopelessly invisible and difficult to maintain. The gears used universally in mechanical clocks wear down over time and offer only approximately correct ratios. A precomputed display requires too many calendar pages, hydraulic calculation requires too much power, and an array of levers requires too slow a timing source. This leaves only *mechanical digital logic*. If the Clock were a large, slow, mechanical computer, it would be as easy to understand as a geared clock—but far more accurate (because error does not accumulate in digital calculation the way it does in an analogue system of gears), and wear would not affect its accuracy.

The question of what temporal information the Clock would display and how such information would be displayed became a

FIGURE 11.1 The first version of the prototype Millennium Clock. Made at half-scale (four feet high) of laser-cut plywood in mid 1998, this model showed how the timing, calculation, and display mechanisms might work together. *Timing* is provided by the torsional pendulum made of two tungsten balls swiveling slowly in the middle. *Calculation* takes place in the five serial bit adder rings at the bottom. Their result is transmitted via the hexagonal geneva wheels to the stacked intercallator gears above the pendulum, which drive the display rings on top. The *display* shows the year (up to 12,000 C.E.) in the outer ring, the Sun position (hence time of day), Moon position and phase, the rising and setting of Sun and Moon, and the locally visible star field (responding to time of day, time of year, and the precession of the equinoxes).

Not shown in either illustration is the *solar trigger*, which keeps the Clock perpetually accurate with an impulse of focused heat from the exact midday sun.

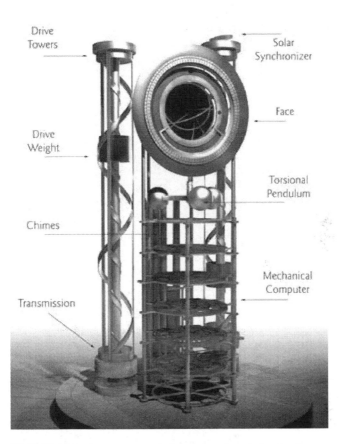

FIGURE 11.2 A computer rendering shows a later design for the prototype, featuring an upright display face at eye level. This illustration includes the *power* source—two columns, each with a weight descending on a helical drive, giving torque to each central shaft. One shaft feeds power to the calculation and display mechanisms, the other provides the incessant tiny nudge that keeps the timing mechanism (three-ball torsional pendulum) rotating. This late 1998 version offers better visibility of the Clock's working. The pendulum rotates at waist level, with its escapement in plain sight, and the five bit adder rings below are wider spaced. The intercallator gears driving the display would be viewable behind the face of the Clock. On top of the right-hand power column is the air fly—an air brake that gentles the motion of the twice-daily calculation process in the adder rings. That reduces wear and tear and makes the process more deliberate and watchable.

matter of much discussion. Each Millennial Clock would probably be somewhat different. Instead of multiple hands and faces Hillis opts for one face and one stationary "hand" with multiple concentric rings rotating under it. The rings could display the orbits of the visible planets (Mercury, Venus, Earth, Mars, Jupiter, and Saturn) and note the date according to the Chinese, Mayan, Jewish, Gregorian, and Islamic calendars. The rings could also predict solar and lunar eclipses for millennia to come.

Yet too much busyness might diffuse the Clock's emotional impact. Gradually the decision emerged to reduce the complexity of the Clock's display. It would show, in briskly intuitive fashion only: the year date (Gregorian calendar, easily convertible to anything else), the position of the Sun (hence the approximate time of day), the Moon's position and phase, the local rising and setting times of the Sun and Moon, and the locally visible star field, which rotates daily, shifts with the yearly seasonal cycle, *and* adjusts very gradually to the 25,784-year cycle of the precession of the equinoxes.

Kevin Kelly observed that such a display would be a return to origins: "From the very beginning clocks were simulacra. The first clocks were models of the heavens—Sun and Moon rotating overhead. Later clockmakers modeled a universe of seasons and time and birth and death, displayed as marching jacks and crowing cocks. Only later, in the minimalist modern period, were clocks abstracted into the naked passage of seconds and minutes." Horologists became obsessed with accurately measuring millionths and then billionths of a second. Hillis headed in the opposite direction.

The Clock's computer is shockingly subtle and simple. As Hillis says, "It's a design Babbage could have succeeded with. A skilled clockmaker could have built one in the fifteenth century." (Charles Babbage's celebrated mid-nineteenth-century mechanical computer, built of brass, failed in part because it used decimal notation. Hillis's mechanical computer is binary.) The subtlety comes with the extreme accuracy that digital computation allows. A calculation thirty-two bits deep is accurate to one in seven million—one day in twenty thousand years. Furthermore, a very-long-period calculation is just as easy as a short one. Computing the glacially slow precessional movement of the star field is no more of a problem than computing the monthly position of the Moon. Contrast this with the heroic effort that an IWC "Da Vinci" mechanical wristwatch has to make to display centuries accurately. The gear train requires a reduction ratio of 6,315,840,000 to 1. After 25.245 billion balance spring oscillations the century display advances by one on the watch face.

In its patent application the Hillis mechanism is called a "Serial Bit Adder": a ring slowly rotating on a disk. Pins are set in the disk to represent a constant (e.g., 29.5305882 days for the Moon's cycle). By moving a simple lever back and forth the ring accumulates the constant in serial bit order—least significant bit to most significant bit. At the precisely appropriate time the accumulator overflows and advances the display ring. Each display ring has its own adder ring. The calculations are made just twice a day, at noon and midnight—very low wear and tear. Journalists have noted the irony—or poetry—of Hillis moving directly from designing and building the world's fastest supercomputers to designing and building the world's slowest computer.

The ingenuity in the Clock is late twentieth century and so are the materials. The process of constructing a prototype Millennial Clock in 1998 led the builders into direct encounters with deep time. The challenge was to build a prototype that not only worked but would keep working for ten thousand years. Exotic materials were sought for maximum longevity and minimal friction: Monel (nickel–copper alloy) and tungsten carbide for hardness, diamond coating for further hardness, Invar alloy for temperature independence, metallic glass for a potentially inexhaustible pendulum ribbon. (For the small prototype Clock the requirements for familiar and not-worth-stealing materials were set aside.)

A typical insight from the work was this note from Long Now engineer Kiersten Muenchinger:

> I gained a somewhat new perspective on designing for 10,000 years last night in a conversation with Rob Ritchey, a professor at Berkeley. Prof. Ritchey was referred to us by the folks at Amorphous Technologies International (metallic glass, Liquid Metal Golf) as a leading researcher in materials science, specifically in fatigue. The spin that Ritchey had was that there has not been enough long-time materials testing on metals to know how much our materials will wear, corrode, or creep. Interestingly, the longest-term data on

materials' properties just came out of Japan a year or two ago and foretells—stretching—only 100 million cycles, where we're looking for on the order of 100 billion cycles. Therefore, using a material that is known to just plain not change state and design completely around that is the way to go. His suggestions were silicon and silicon carbide. I expressed my reservations as to the brittleness of these materials and he said yes, of course they're brittle, but we'll know whether a Si part works for 10,000 years because it will catastrophically fail immediately if it doesn't work. If it doesn't fail on you immediately, well then there you are!

Another approach of course is to stockpile spare parts for gradual replacement over the millennia. Contemplating the heaps of backup components might be part of visiting the Clock. Each of the successive Clocks is meant to be a more immersive experience. The prototype is eight feet tall, a city Clock is likely to be twenty feet high, and a desert Clock is hardly worth doing if it is less than sixty feet high. (There is a technical term for this: "Within the human cortex crouches an impulse to build something huge," declares a book about the enormous Indian ruins at Chaco Canyon. "Anthropologists call that sudden urge, when acted on, a florescence.")

At 4 A.M. one day in 1997, Danny Hillis woke up with a picture in his mind. He wrote immediately to his colleagues:

Imagine the Clock is a series of rooms. In the first chamber is a large, slow pendulum. This is your heart beating, but slower. In the next chamber is a simple 24-hour clock that goes around once a day. In the next chamber, just a Moon globe, showing the phase of the lunar month. In the next chamber is an armillary sphere tracking the equinoxes, the solstices, and the inclination of the Sun. The sphere also contains the Sun-tracking lens that synchronizes the Clock. This room also contains the weights, which must be wound each year. The next chamber is the lifetime room—a single blank, featureless disk of soft stone that rotates once a lifetime, onto which you can carve your own mark.

The final chamber is much larger than the rest. This is the calendar room. It contains a ring that rotates once a century and the 10,000-year segment of a much larger ring that rotates once every precession of the equinoxes. These two rings intersect to show the current calendrical date. This is also the room that holds the mechanism of the Clock. Maybe now that I have written this down I can get back to sleep. I hope I still like it when I wake up. Good night.

His fellow board members were enthusiastic but ventured that perhaps Danny had been working at Disney too long. His dream sounded suspiciously like a theme park ride.

"Time *is* a ride," Hillis replied, "and you are on it."

BURNING LIBRARIES

In Tom Stoppard's contemporary play, *Arcadia*, the budding genius Thomasina laments to her tutor about the burning of the ancient Library of Alexandria, "Oh, Septimus!—can you bear it? All the lost plays of the Athenians! Two hundred at least by Aeschylus, Sophocles, Euripides—thousands of poems—Aristotle's own library! . . . How can we sleep for grief?"

The burning of libraries is so universally regarded as a crime against humanity that it is instructive to examine the historical record. What exactly is the outrage about?

The legendary stature of the Library of Alexandria is justifiable. In its prime (290 B.C.E. to 88 B.C.E.) the library was the fount of the Hellenistic Greek Renaissance, and the classics it preserved helped inspire the fifteenth-century European Renaissance. Then the largest city in the world, Alexandria remained the intellectual capital of the Mediterranean for most of the duration of the Roman Empire. The famous library and museum at its peak may have held six hundred thousand scrolls—the equivalent of one hundred twenty thousand modern books. Alexandria's library was an intensely productive community of writers, translators, editors, historians, mathematicians, astronomers, geographers, and physicians. Its librarians included Apollonius of Rhodes (poet of *The Argonauts*), Callimachus (the father of bibliography), Eratosthenes (who estimated the diameter of the Earth), Aristarchus of Samos (a Sun-centered Copernican eighteen centuries before Copernicus), and Hipparchus (discoverer of the precession of the equinoxes). By dint of exhaustive collection and close scholarship, canonical editions of classics such as Homer, Plato, and the Athenian playwrights were created and distributed. Later, the Hebrew Bible was translated into Greek in Alexandria. This African city was directly responsible for much of what Europe became.

Who burned the Library of Alexandria? Ptolemy VIII torched much of the city in a civil war in 88 B.C.E. and scattered the scholars, at least temporarily. In 47 B.C.E. Julius Caesar, escaping assassination, set fire to the Alexandrian fleet, which set fire to parts of the city, including buildings containing forty thousand scrolls. In 273 C.E. the Roman Emperor Aurelian reconquered Egypt, burning the part of Alexandria where the library was. In 391 C.E. the Christian Archbishop Theophilus deliberately incinerated Alexandria's "daughter library" of forty thousand scrolls because it was housed in a pagan temple of Serapis. In 645 C.E. the Muslim conqueror Caliph Omar answered one of his generals who inquired what to do with Alexandria's renowned books: "If their content is in accordance with the book of Allah, we may do without them, for in that case the book of Allah more than suffices. If, on the other hand, they contain matter not in accordance with the book of Allah, there can be no need to preserve them. Proceed, then, and destroy them." The chronicler Ibn al-Kifti reports that the books were used to heat the public baths of Alexandria: "it took six months to burn all that mass of material."

So who burned the Library of Alexandria? War did three times, inadvertently. Religious bigotry did twice, on purpose. We are right to grieve. Only one in ten of the major Greek classics survived. Nothing like Alexandria's library was seen again for a thousand years.

A different reason for burning books was originally invented by China's first great emperor, Shih Huang-ti, in the third century B.C.E. His Ch'in dynasty provided China with its name, its approximate boundaries, and the totalitarian form of administration that would recur for two millennia. He built the Great Wall and standardized the measuring systems and written language of China. His twenty-square-mile burial complex included an entire terra cotta army (only recently rediscovered). His dynasty, he proclaimed, would "make the world a single unity for ten thousand generations."

In 213 B.C.E. at an imperial banquet a Confucian scholar offered criticism of such a severe break with the past. "Nothing can endure for long," he said, "but that which is modeled on antiquity." The emperor's grand councilor Li Ssu responded, "There are some men of letters who do not model themselves on the present, but study the past in order to criticize the present age. They confuse and excite the ordinary people. If such conditions are not prohibited, the imperial power will decline above and partisanship will form below." The order went out to burn *all* books in the empire except only those dealing with agriculture, medicine, and fortune telling. To even discuss the forbidden works was punishable by death. The pathetically few early classics to survive the conflagration had to be later rewritten from memory.

Shih Huang-ti had set about simply erasing the past in order to ensure his version of the future. Instead, his ten-thousand-generation dynasty ended shortly after his death, having lasted fifteen years (221–206 B.C.E.). The Confucian Han dynasty that followed vilified Shih Huang-ti, and he remains largely a villain in Chinese history. Yet this did not prevent Mao Tse-tung from attempting a similar erasure with his Cultural Revolution in 1966, to disastrous effect.

The same impulse inspired Hitler's book-burning ceremonies of May 1933. "The German form of life is definitely determined for the next thousand years," he declared. "There will be no other revolution in Germany for the next thousand years." In a square by the University of Berlin twenty thousand books were burned. Other cities did the same. Into the flames went the "subversive" works of Thomas Mann, Erich Maria Remarque, Albert Einstein, Hugo Preuss, Sigmund Freud, Marcel Proust, André Gide, Émil Zola, H. G. Wells, and Jack London. The new propaganda minister, Joseph Goebbels, told the students at the bonfires, "These flames not only illuminate the final end of an old era; they also light up the new." Henceforth all cultural activity would be regimented by the state. Shorter even than the Ch'in dynasty, Hitler's thousand-year Reich lasted just over twelve years, and humanity remembers him as a monster.

Starting anew with a clean slate has been one of the most harmful ideas in history. It treats previous knowledge as an impediment and imagines that only present knowledge deployed in theoretical purity can make real the wondrous new vision. Thus the French Revolution of 1789, the Russian Revolution of 1917, and the Chinese Communist Revolution of 1949 each made brave new worlds that catastrophically failed. By cutting off continuity with the slower parts of their cultures they had no fallback. The American Revolution of 1776, by contrast, was highly conservative. Its instigators studied Roman, Venetian, and even Iroquois history for precedents. There was little of the brutal rhetoric of making a total break with the past. As a result, all the leaders who started the revolution lived to see it through to completion, and its innovations in governance aged relatively well. The Americans severed the political bonds with the Old World but not the cultural bonds. They burned their bridges, not their libraries.

A lethal variant of religious bigotry was at work in the sixteenth century when the Spanish Conquest missionaries burned the codices of the Mayans. The most intellectual of the American peoples, the Maya had a written language, advanced mathematics (complete with a zero), and a more precise calendar system than any in Europe. Of their thousands of bark-paper books of history, song, myth, astronomy, genealogy, and prophecy, only four survived the flames. This was deliberate cultural genocide of one people by another. The lead cleanser of pagans was the Franciscan friar Diego de Landa. He studied Mayan culture carefully, the better to destroy it. Coming across a trove of thirty beautiful hieroglyphic codices, he reported that since "they contained nothing in which there was not to be seen superstitions and falsehoods of the devil, we burned them all, which [the Mayans] took most grievously, and which gave them great pain." Now that archaeologists have learned from hieroglyphs on buildings and pottery how to read Mayan text, the pain continues indefinitely.

The danger also continues. Cultural arson was unleashed again in August 1992, in Bosnia. For three days Serb forces targeted Sarajevo's multicultural National and University Library with a bombardment of incendiary grenades. Bosnia's written heritage was consumed—a million and a half volumes, one hundred fifty-five thousand of them rare books and manuscripts. The library's director said that the Bosnian Serbs "knew that if they wanted to destroy this multiethnic society, they would have to destroy the library." In horror at the event, librarians worldwide have established an Internet-based Bosnian Manuscript Ingathering Project to track down duplicates and replace as many as possible of the lost documents.

Burning libraries is a profound form of murder, or if self-inflicted, suicide. It does to cultural continuity—and hence safety—what destroying species and habitats does to nature's continuity, and hence safety. Burning the Amazon rain forest burns the world's richest library of species. *The accumulated past is life's best resource for innovation.* Revolutions cut off the past. Evolution shamelessly, lazily repurposes the past. Reinventing beats inventing nearly every time.

DEAD
HAND

Respect for the past can be overdone. Too much heritage-fixation becomes a form of self-obsession that makes one's heritage convoluted and therefore decadent. Deep creativity becomes oppressed and the young are poisoned with ancient hatreds.

The following is an account of how public water became foul in nineteenth-century European cities:

> Christians preferred to be buried in the sanctified ground of the graveyard, creating layer upon layer of putrefying parishioners, all leaching down to the groundwater. Over time, urban churches had laid hundreds of thousands of believers to rest in a small area. The Scottish journalist Basil Hall writes in 1820 of churchyards ankle deep with "rank and offensive mould, mixed with broken bones and fragments of coffins," while the water of city wells near churches grew redolent with putrescine, cadaverine, and other pungent organic compounds associated with decomposition.

As a Californian reading the passage, I thought, "That's Europe, all right." Of course I adore Europe, and like all tourists I complain when anything new clutters up the glorious ancient cityscapes and landscapes, but I would never go there to start anything.

Radical revolution makes perfect sense in overcivilized Europe. How could a Luther or Marat, Cromwell or Lenin *not* want to bring down the rotten, towering edifice of the past? Make a clean sweep! What is indeed excessive continuity inverts into deeply destructive loss of continuity: The stained-glass windows are shattered along with the ossified clergy.

This Europe–America dialogue is an old one and it is about something real. In 1856 the American novelist Nathaniel Hawthorne, suffocated by a visit to the British Museum, wrote, "I do not see how future ages are to stagger onward under all this dead weight, with the additions that will continually be made to it." The American ambassador to London, journalist Walter Hines Page, advised more quotably, "Fling off the dead hand of the past!"

Looking in the other direction, Goethe wrote, "America, you have it better than our old continent; you have no ruined castles and no primordial stones. Your soul, your inner life remain untroubled by useless memory and wasted strife." Other Europeans, to be sure, are at pains to remind Americans constantly (and correctly) how shallow, callow, and fevered they are.

The Europe–America dialogue is not a problem; it is one form of solution. While Europe specializes in deep continuity with occasional equally deep discontinuity, America specializes in perpetual petty turmoil. America provides the stimulation in the arrangement, Europe provides the wisdom. America is comic, Europe tragic. Together they make great theater. The *kairos-chronos* dialogue can be between different parts of the world; it can be between different parts of the mind. The one contrives, the other warns.

Are there other solutions that might work *within* a culture or an institution? As the world increasingly globalizes, we cannot count on always having different regions working in different time frames, though this probably should be defended for as long as possible.

One highly evolved solution is to compartmentalize the past. Park it in academic disciplines. ("Writing history is a way of getting rid of the past"—Goethe again.) Park it in august buildings. ("Layer upon layer, past times preserve themselves in the city until life itself is finally threatened with suffocation; then, in sheer defense, modern man invents the museum"—Lewis Mumford.) Park the past in literature and in theater. Park it in religion. (Yet why are Europe's beautiful cathedrals empty while America's storefront churches are full?)

A new version of the problem is coming. As Hawthorne predicted, the amount of accumulated past is accelerating. Each new U.S. president leaves behind more papers to be preserved than all the previous presidents combined. At the same time, with digital media it is increasingly possible to store absolutely everything. The traditional role of the librarian and curator—to select what is to be preserved and ruthlessly weed out everything else—suddenly is obsolete.

How do we keep from being crushed by the total past, instantly accessible on the Net? Brian Eno observes, "I had a large house once. Effectively it had infinite cheap storage. I've never been so miserable as when I found myself living among the unselected heaps of crap that I'd accumulated. I favor savage selection, but everyone making their own. That way you get a myriad of perspectives instead of one and instead of an undifferentiated heap." This pretty well describes how the World Wide Web is organizing itself.

People complain about overwhelming masses of information on the Web, but one of its inventors, Tim Berners-Lee, comments, "To be overloaded by the existence of so much on the Web is like being overloaded by the mass of a beautiful countryside. You don't have to visit it, but it's nice to know it's there. Especially the variety and freedom."

The Internet may be showing the way to live with an infinite amount of past in infinite detail, and still encourage freedom to innovate without the need of violent revolution. Add in drastic life extension, due soon, and you get quite a different world, one that might say about our world: "Can you imagine what it was like when people and programs had to die, whether they wanted to or not? No wonder it took so long for culture to get anywhere. Everything and everyone was starting over all the time—the same dumb mistakes over and over again. People were either pastless or trapped in the past. Their lives were as beautiful and tragic and stupid as waves breaking on the beach."

ENDING
THE DIGITAL
DARK AGE

The promise has been made. "Digital information is forever. It doesn't deteriorate and requires little in the way of material media." So said one of the chieftains of the emerging digital age, computer chip maker Andy Grove, head of Intel Corporation. Another chieftain, Librarian of Congress James Billington, has worthily set about digitizing the world's largest library so that its contents can become accessible by anyone, from anywhere, forever.*

But a shadow has fallen. "It is only slightly facetious," wrote RAND researcher Jeff Rothenberg in *Scientific American*, "to say that digital information lasts forever—or five years, whichever comes first."

Digitized media do have some attributes of immortality. They possess great clarity (a bit can only be a zero or a one); great universality (binary notation conquered the world without a fight); great reliability (any file can be error-checked to ensure that it is intact in every detail); and great economy (digital storage is already so compact and cheap that it is essentially free). We are still getting used to this. Many people have found themselves surprised and embarrassed by the reemergence of perfectly preserved E-mail or online newsgroup comments they wrote nonchalantly years ago and forgot about.

Yet those same people discover that they cannot revisit their own word processor files or computerized financial records from ten years earlier. It turns out that what was so carefully stored was

---

*The genesis of this chapter was a conference held at The Getty Center in Los Angeles in February 1998. "Time & Bits: Managing Digital Continuity" was sponsored by the Getty Conservation Institute, the Getty Information Institute, and The Long Now Foundation. Participants were Peter Lyman, Howard Besser, Danny Hillis, Brewster Kahle, Jaron Lanier, Doug Carlston, Kevin Kelly, Brian Eno, Stewart Brand, Margaret MacLean, and Ben Davis. A book of the proceedings is available from The Getty Center, 1200 Getty Center Drive, Los Angeles, CA 90049.

written with a now-obsolete application, in a now-obsolete operating system, on a long-vanished make of computer, using a now-antique storage medium (where do you find a drive for a 5 1/4-inch floppy disk?). It doesn't matter that in those days everyone wrote with WordStar, in CP/M, on a Kaypro computer. No one does now, and almost no one can.

If you write something this week with Word in Windows 98 on a Dell computer, what are the chances of anybody being able to read it in 2008? The same doubt hangs over the big iron—the mainframes and minicomputers that process the digits that run and record our world.

It doesn't matter how widely used the machines are. Jaron Lanier, the inventor of immersion technologies called *virtual reality*, recently reported,

> I was asked last year by a museum to display an art video game ("Moondust") that I had written in 1982. It ran on a Commodore 64, a computer that had already sold in the millions by the time of the game's release. It turns out that after my game cartridge was introduced, there was a slight hardware change to the computer (in 1983), which caused the sound to not work. So I had to find a 1982 Commodore 64. But then it turned out that all the joysticks I could find only worked on the later version. Once I finally had a matching trio of computer, joystick, and cartridge, it turned out that I didn't have a working video interface box. All this trouble with a machine whose operating system was fixed in ROM and had been available at the time in the millions!

After months of effort, Lanier gave up.

Fixing digital discontinuity sounds like exactly the kind of problem that fast-moving computer technology should be able to solve; but it can't, because fast-moving computer technology *is* the problem. By constantly accelerating its own capabilities (making faster, cheaper, sharper tools that make ever faster, cheaper, sharper tools) the technology is just as constantly self-obsolescing. The great creator is the great eraser.

Behind every hot new working computer is a trail of bodies—of extinct computers, extinct storage media, extinct applications, extinct files. The science-fiction writer Bruce Sterling refers to our time as "the Golden Age of dead media, most of them with the working life span of a pack of Twinkies." On the Internet Sterling is mustering a roll call of their once-honored names. Just among personal computers are Altair, Amiga, Amstrad, Apples I, II, III, Apple Lisa, Apricot, Atari, AT&T, Commodore, CompuPro, Cromemco, Epson, Franklin, Grid, IBM PCjr, IBM XT, Kaypro, Morrow, NEC PC–8081, Northstar, Osborne, Sinclair, Tandy, Wang, Xerox Star, Yamaha CX5M. Buried with them are whole clans of programming languages, operating systems, storage formats, and countless rotting applications in an infinite variety of mutually incompatible versions. Everything written on them was written on the wind, leaving not a trace.

In personal terms the loss seems merely an inconvenience, the price of progress, but in terms of civilization the loss is catastrophic. Just when we think at last, thanks to digitization, everything we want to keep can be preserved perfectly forever, the reality is precisely the opposite. Never has there been a time of such drastic and irretrievable information loss. If this seems extravagant, consider the number of literate people in today's world, the bulk of whose work is knowledge-based, which means increasingly relying on computers. The world economy itself has become digital.

Danny Hillis notes that from previous ages we have good raw data written on clay, on stone, on parchment and paper, but from the 1950s to the present recorded information increasingly disappears into a digital gap. Historians will consider this a dark age. "For example," Hillis recalls, "when we finally shut down the old PDP–10 [pioneer minicomputer] at the MIT Artificial Intelligence Lab, there was no place to put files except onto mag tapes that are by now unreadable. So we lost the world's first text editor, the first vision and language programs, and the early correspondence of the founders of artificial intelligence." Science historians can read Galileo's technical correspondence from the 1590s but not Marvin Minsky's from the 1960s.

Not only do file formats quickly become obsolete, the physical media themselves are short lived. Magnetic media such as disks and tape lose their integrity in five to ten years. Optically etched media such as CD-ROMs, if used at all, only last five to fifteen years before they degrade; and digital files do not degrade gracefully like analog audio tapes. When they fail, they fail utterly. You can't open them. They have one-tenth the readable life span of acid-laced newsprint.

Beyond the evanescence of data formats and digital storage media is a still deeper problem. Large-scale computer systems are at the core of driving corporations, public institutions, and indeed whole sectors of the economy: financial markets, utilities, telecommunications, travel and distribution, health care, and government. Over time these gargantuan systems become dauntingly complex and unknowable as new features are added, old bugs are worked around with layers of "patches," generations of programmers add new programming tools and styles, and portions of the system are repurposed to take on novel functions. With a mixture of respect and loathing, computer professionals call these monsters *legacy systems*.

Typically, outdated legacy systems make themselves so essential over the years that no one can contemplate the prolonged trauma of replacing them, and they cannot be fixed completely because the problems are too complexly embedded and there is no one left who understands the whole system. Teasing a new function out of a legacy system is not done by command but by conducting a series of cautious experiments that with luck might converge toward the desired outcome.

Here's the real fear. Thanks to proliferating optical-fiber land lines worldwide and the arrival of low-Earth-orbit data satellite systems such as Teledesic, we are in the process of building one vast global computer. ("The network is the computer," proclaims Sun Microsystems.) This world computer could easily become the Legacy System from Hell that holds civilization hostage: The system doesn't really work, it can't be fixed, no one understands it, no one is in charge of it, it can't be lived without, and it gets worse every year.

So far computers are very bad at dealing with time. Society may wind up grateful for the chaos and confusion caused by the Year 2000 problem, when countless computer systems still laced with chips, code, and files that use two-digit year dates choke on the year date "00." All the right warnings are given by the event. We are forced to realize that software thus far is brittle, and that a software-obligate civilization inherits that brittleness; it can crash. Programmers and their managers still don't realize that the code they write and conventions they adopt are more likely to be embedded than replaced; today's bleeding-edge technology is tomorrow's broken legacy system. Commercial software is almost always written in great haste, at ever-accelerating market velocity; it can foresee an "upgrade path" to next year's version, but decades are outside its scope. And societies live by decades, civilizations by centuries.

So far, the mismatch between computers and civilization is growing.

We probably cannot and should not reverse the digitization of everything. What we can do is convert the design of software from brittle to resilient, from heedlessly headlong to responsible, and from time corrupted to time embracing. Achieving each of these qualities is known to be an approximately intractable problem. To be sure none can be solved in a year, but all can yield to decades of focused work if we understand that the health of civilization is at stake.

The toughest problem will be building software that will forgive us our trespasses and trespass not against us. There is no real glimmer yet how to approach the question but also no reason to give up on it. As for computer professionals routinely thinking and acting with long-term responsibility, that may come gradually as a by-product of the Year 2000 comeuppance, of life extension, of environmental lessons, and of globalization (island Earth). At issue here is how to address the management of digital continuity over time, how to shorten the digital dark age.

Pure information can have astonishing longevity. In 1090 C.E. the Chinese genius Su Sung built a monumental water-driven mechanical clock for his emperor. It was dazzling, two centuries ahead of anything like it in Europe. Yet the succeeding emperor discredited it, vandals made off with the bronze parts, and the clock and all its ingenious devices were utterly forgotten. In the nineteenth century an illustrated manuscript, lost since 1172, turned up: "New Design for a Mechanized Armillary Sphere and Celestial Globe," by Su Sung. The description is so complete that working replicas of the original clock have been built. The material clock lasted only a decade; the informational clock, indefinitely.

How much information exists in the world today, and how does it compare with digital storage capabilities? Michael Lesk at Bellcore has done the detailed calculation. The Library of Congress is routinely credited as containing twenty million books, hence twenty terabytes of text information (that would fill twenty thousand hundred-megabyte Zip disks, to use a quaint 1998 standard). Counting all of its graphic, film, and sound archives, the Library of Congress grand total multiplies fifteen hundredfold to three petabytes, or thirty million Zip disks. (One gigabyte = one thousand megabytes; one terabyte = one thousand gigabytes; one petabyte = one thousand terabytes.) Lesk estimates that the total digital content of the World Wide Web surpassed what is in the Library of Congress in 1998, and it keeps doubling every few months.

The comprehensive total of information in the world—counting every picture postcard, phone call, web link, and television commercial—Lesk estimates at roughly twelve thousand petabytes: one hundred twenty billion Zip disks, four thousand times the Library of Congress. In 1998 a major (but unheralded) milestone occurred: Available digital data storage capacity surpassed the total of information in the world. We now have more room to store stuff than there is stuff to store. In other words, concludes Lesk, "We will be able to save everything—no information will have to be thrown out—and the typical piece of information will never be looked at by a human being." Most information will simply be exchanged among computers. Brewster Kahle's Internet Archive is attempting to download and preserve the entire World Wide Web. The easy part of that Herculean endeavor is the digital storage.

Such a deluge of data, accelerating every month, does bring its own problems. The vast archives of digitized NASA satellite imagery of the Earth in the 1960s and 1970s—priceless to scientists studying change over time—now reside in obsolete, unreadable formats on magnetic tape. Will NASA find the money and time to translate all that data to current media before the tapes decay, while new satellite data keeps flooding in? It's doubtful. Physicist Neil Gershenfeld at MIT's Media Lab worries that handling the constant arrival of new bits will keep us from ever managing the old bits properly. Loss of cultural memory becomes the price of staying perfectly current.

If raw data can be kept accessible as well as stored, history will become a different discipline, closer to a science, because it can use marketers' data-mining techniques to detect patterns hidden in the data. You could fast-forward history, cross-correlate freely, zoom in on particular moments. Watershed events might be studied in the original: the actual force-feedback virtual-reality experiment that showed a new way to fold a protein that transformed medicine, plus the lab surveillance camera images of the event, as well as the phone calls, E-mail, and Web searches that surrounded the discovery.

Note the two different kinds of digital records in this example: passive and active. The E-mail, phone calls, and photographs are passive; all you have to do is keep them readable. The virtual reality experiment, however, is active; it was probably run on some experimental one-off piece of lab equipment cobbled together in a fragile array of then-current tools. Without that complex of hardware, you can't replay the experiment. Preservation of such hardware-dependent digital experiences is nearly impossible, says Jaron Lanier. For instance, an elaborate virtual-reality model of Berlin has been used for planning the city for years, but this invaluable artifact will almost certainly be lost eventually. The U.S. Army's famous computer model of the pivotal tank battle in the Gulf War, refought by countless soldiers in the years following the war, is likewise doomed in its original form.

Digital storage is easy; digital preservation is hard. Preservation means keeping the stored information catalogued, accessible, and usable on current media, which requires constant effort and expense. "Back when information was hard to copy, people valued the copies and took care of them," says Danny Hillis. "Now, copies are so common as to be considered worthless, and very little attention is given to preserving them over the long term." Furthermore, though contemporary information has economic value and pays its way, there is no business case for archives, so the creators or original collectors of digital information rarely have the incentive—or skills or continuity—to preserve their material. It's a task for long-lived nonprofit organizations such as libraries, universities, and government agencies, who may or may not have the mandate and funding to do the job.

Migration is the term digital archivists use for the transfer of files from one computer platform and generation of applications to another—from VisiCalc on an Apple II, say, to Excel on a PC, to whatever comes next. Some archivists claim that the need for migration is good because it forces the habit of steady attention that preservation requires, but most worry that the process is so difficult that even a temporary loss of interest, funding, or competence can break the chain of migration, and all the previous work is lost.

Digital archivists thus join an ancient lineage of copyists and translators reaching back through European monastic scribes to the Hellenistic scholars at the Library of Alexandria. The process, now as then, can introduce copying errors and spurious "improvements" and can lose the equivalent of volumes of Aristotle; yet the practice also builds the bridge between language eras, from Greek to Latin to English to whatever follows. I think that to become comfortable about digital continuity—to feel assured that our future will stay connected to our past, that the digital dark age is ending—we will need the framework for a universal translation system. What might that mean?

The test will be having the tools to bridge long gaps in the chain of migration. The materials that historians and scientists most treasure often were unvalued originally. How can researchers reach back through five or fifty generations of high technology to revive a file or program that turned out to be seminal, or revealingly typical, or a crucial data trove?

The archivist Howard Besser points out that digital artifacts are increasingly complex to revive. For starters you've got the viewing problem: a book displays itself, but the contents of a CD-ROM are invisible until opened on something. Then there's the scrambling problem: the innumerable ways in which files are compressed and, increasingly, encrypted. There are interrelationship problems: hypertext or web-site links active in the original that now are dead ends. And translation problems occur in the way different media behave: Just as a photograph of a painting is not the same experience as the painting, looking through a screen is not the same as experiencing an immersion medium; watching a game is not the same as playing the game.

For all these reasons archivists now encourage tagging all digital artifacts with a rich supply of *metadata*—that is, digital information about the artifact identifying what it is and how it works. A number of professional organizations are working on setting consistent (and improvable) standards for metadata. A set of "best practices" gradually is emerging for ensuring digital continuity: Use the most common file formats, avoid compression where possible, keep a log of changes to a file, employ standard metadata, make multiple copies, and so forth. And don't forget atomic backup: While the durability of bits is still moot, the atoms in ink on paper have great stability. As archivist Besser writes, "The default condition of paper is persistence, if not interrupted; the default condition of electronic signals is interruption, if not periodically renewed."

Another approach is through core standards, such as the DNA code in genes or written Chinese in Asia, both readable through epochs while everything changes around them and through them. The platform-independent programming language called Java boasts the motto, "Write once, run anywhere." One of Java's creators, Bill Joy, asserts that the language "is so well specified that if you write a simple version of Java in Java, it becomes a Rosetta Stone. Aliens, or a sufficiently smart human, could eventually figure it out because it's an implementation of itself." In other words, "Write once, run anytime." We'll see.

Exercise is the best preserver. Jaron Lanier notes that documents such as the Torah, the Koran, and the I Ching are impressively persistent because every age copies, analyzes, critiques, and uses them. The books live and are kept contemporary by use. Since digital artifacts are rapidly outnumbering all possible human users, Lanier recommends employing artificial intelligences to keep the artifacts exercised through decades and centuries of forced contemporaneity, kept ever up to date for a potential human user.

Still, even robot users might break continuity. Most reliable of all would be a two-path strategy suggested by Doug Carlston, cofounder of Brøderbund Software. One path is slow, periodic, and conservative; the other is fast, constant, and adaptive. To keep a digital artifact perpetually accessible, record the current version of it on a physically permanent medium such as microetched silicon disks, then go ahead and let users, robot or human, migrate the artifact through generations of versions and platforms, pausing from time to time to record the new manifestation on a silicon disk. Over time the two paths keep each other reliable and current. When the chain of use is eventually broken, it leaves behind a permanent record of the chain up to that time, so the artifact can be brought back to life for researchers or a new generation of users to begin the chain anew.

And there is the Net. Just weeks after Jaron Lanier gave up on reviving Moondust, his pioneer computer game, he got E-mail from someone named Walter, who told him,

> Time disappeared today. Soft, lazy swirls of color danced on my Power Mac monitor like underwater synchronized swimmers, danced to spacy, otherworldly music. Once again, fifteen human years later, techno-eons later, "Moondust" fell softly on my delighted eyes and ears. . . . It took me about three hours of Web detective work to find it. There's an amazing amount of time and energy devoted to technostalgia out there. Emulators. Atari 2600 emulators, Intellivision emulators, even Vic–20 emulators. . . . Not only did I find "Moondust," I found the "Moondust" documentation.

The Internet is populated with legions of amateur digital archivists, archaeologists, and resurrectionists. They track down the original code for lost treasures such as Space Invaders (1978), Pac-Man (1980), and Frogger (1981), and they collaborate on devising emulation software that lets the primordial programs play on contemporary machines. The emulation techniques pioneered by such vernacular programmers are at present the most promising path to a long-term platform migration solution.

Such massively distributed research can convene enormous power. For instance, thanks to the current interest in genealogy, the thousands of users of a program called Family Tree Maker are linking their research into a World Family Tree on the Web. So far it has tied together seventy-five thousand family trees, a total of fifty million names. The goal, once unthinkable, is to eventually document and link every named human who ever lived.

With the Net, preservation goes fractal: infinitely branched instead of centralized. Yet this leaves the question, Is the Net itself profoundly robust and immortal, or is it the most ephemeral digital artifact of all?

The problem of the acceleration of high technology is that its headlong urgency routinely displaces what is important in the long term. Digital industries must shift from being the main source of society's ever-shortening attention span to becoming a reliable guarantor of long-term perspective. We'll know the shift has happened when programmers begin to anticipate the Year 10,000 problem and assign five digits instead of four to year dates. "02002," they'll write, at first frivolously, then seriously.

10,000-YEAR LIBRARY

Taking ten thousand years of future seriously is an interesting undertaking. For instance, what on Earth would something aspiring to be a 10,000-Year Library be good for? One answer might be that it would provide, and even embody, the long view of things, where responsibility is said to reside. Another would be that such a library could conserve the information that is needed from time to time for the deep renewals of renaissance. These are traditional library justifications. The added element is that ten thousand years is an extremely long view, a period in which there are likely to be profound cataclysms requiring many-leveled renewal. Building real value into a 10,000-Year Library could be an intellectual adventure as challenging as space travel.

One very contemporary reason is to make the world safe for rapid change. A conspicuously durable Library gives assurance: *Fear not. Everything that might need to be remembered is being collected and stashed, easily accessible but out of your way. Innovate as intensely as you want. If we head down a blind alley, or get lost, we can pick up the prior path. And we're always free to mine the past for good ideas.* The U.S. Library of Congress is referred to as a "strategic information reserve" by its current head, James Billington. In an increasingly knowledge-oriented world economy that's valid. Beyond these reasons, another glimmers; I'll come back to it.

Thus far The Long Now Foundation's 10,000-Year Library has been an occasion for brainstorming. I'll float here some of the ideas that have turned up in the early pondering, looking first at long-view issues and then at deep record keeping. You're invited to add to these ideas, or even better, forge ahead and implement any that appeal to you via the Net, some existing library, or whatever else is handy. A cultural Darwinian once declared, "What has been done, thought, written, or spoken is not culture; culture is only that fraction that is remembered." Such formal remembering used to be the province strictly of elites. No longer.

In terms of stepping bodily into the long view the 10,000-Year Library should be a physical place. Fantasy immediately calls up a refuge from the present, a place of weathered stone walls and labyrinthine stacks of books, at a remote location with far horizons. It is a place for contemplative research and small, immersive conferences on topics of centenary and millennial scope. In a timeless reading room it would be wonderful to have a collection in which every volume you lay your hand on, on any subject, is superb—selected by committees of specialists of great probity and judgment. You should be able to easily buy or have personally printed a copy of any book that wins your interest.

The fantasy continues: What might be the best time-spanning, future-engaging categories to collect? History, obviously, and historiography: the history of the idea of history. Archaeology and paleontology, for the long human perspective. Environmental books, for their reach into the future. Science fiction, for the same reason, organized by date rather than author, so the browser could scan the progress of the zeitgeist about the future. (The world's best in science fiction is the Eaton Collection of some four hundred thousand volumes at the University of California Riverside Library.) Likewise, nonfiction books about the future. Science and technology books, because their subject has become a major driver of history and is likely to remain so. Demographic and epidemiological texts, for trend analysis. And sundry Long Now special interests: texts on libraries, clocks, and durable institutions. A library such as this, in token form, produced the book you are now reading.

Another image of the 10,000-Year Library is of a vast underground complex hewn out of rock—preferably a mountain, so some of the tunnels have a view. There the Library might be somewhat safe from increasing surface land values, earthquakes, erosion, cosmic rays (harmful to digital media), and the random destruction of warfare. An underground Library offers great mythic potential as well. Such places usually are secret (government) or dangerous (mines, caves), but they have no reason to be either. They can be extremely delightful—for example, the old underground limestone quarry at Les Baux, France, now a tourist attraction. In recent years underground excavation and construction have become as cheap as surface construction.

Escape from the present is also escape from relevance. The emergence of the Net is so much the dominant event of our time that an ambitious new library has no choice but to either try to get out in front of it or utterly turn away from it. I think Long Now's Library should do both: ride the shock wave front *and* its bounce. It is said that the structure of archives always mirrors the structure of power. With the Net this formula is reversed; power is now defined by the grassroots-driven Net structure of archives. Citizens can decide to cross-correlate databases of atrocity victims and military service records to track down specific government-sanctioned torturers and assassins, as Patrick Ball has done with archives in Salvador, Guatemala, and South Africa, working from his base in Washington, D.C. Citizens soon will be able to directly access government records, such as the five billion documents stored with the U.S. National Archives. Via the Net, citizens now correlate the voting records of senators and members of Congress with the sources of their campaign funding, and publicize what they find.

The structure of power used to be the structure of successful lying. In a hierarchy you can lie and make it stick, because higher authority will back up the lie and punish whistle-blowers. Lying is far more flagrant on the Net, but no one can make it stick, because anyone can challenge the lie directly and make their case with multiple links to corroborating sources. One such man, Ken McVay, undermined all the online Holocaust-denying discussion groups in this fashion, connecting them to a linked distributed archive of documentation proving that the Holocaust indeed took place.

Metcalfe's Law of exponential growth of the Net is proving to be even more significant than Moore's Law of exponential growth of microchip capability. The chip is an individual's tool; the Net is society's tool. It may even become its own tool. As the science-fiction writer Vernor Vinge has suggested, the Net is supplied with so much computer power and is gaining so much massively parallel amplification of that power by its burgeoning connectivity that it might one day "wake up." Brewster Kahle, of the Internet Archive, asks, "What happens when the library of human knowledge can process what it knows and provide advice?"

At the same time Long Now is contemplating a timeless desert retreat it has to explore how it can foster on the Net the types of services monasteries provided to deurbanized Europe after the fall of Rome and that universities provided to cities after the twelfth century. Every potential service of the Library therefore should be examined in terms of how it might develop at Net velocity and how it might be something timelessly physical—and how both forms might enhance one another.

For instance, as a way for people to take the future personally, Doug Carlston has suggested that the Library provide mail service through time. According to futurist Richard Slaughter, "It is an unusually moving thing to initiate a message which will not be read until long after one's death. It concentrates the mind effectively. In such a message, one speaks from the heart, is keenly aware of passing time and is also deeply aware of the implicit presence of future people."

Time mail could be started on the Net, but it would be more impressive as a physical experience. At a Clock site you could peruse other people's messages to the future (those that they have okayed for public reading), ponder what you want to say, to when, and perhaps to whom. Then have your message inscribed on titanium (or whatever), pay postage proportionate to the time you want to span, deposit your message in the appropriate mail slot, and watch it slide into the appropriate time capsule.

Time capsules, by the way, are a splendid and common future-oriented practice—hundreds of thousands have been buried—yet some 70 percent are completely lost track of almost immediately. The Library might offer a registry service for time capsules, remembering when they are supposed to be opened and providing maps to people currently at the site. According to Kevin Kelly, who studies time capsules, the most effective are opened periodically, enjoyed, then sealed again with new artifacts added each time.

The most impressive time capsule project I know of is headquartered in Tokyo. The Biological and Environmental Specimen Time Capsule 2001 team hopes to bury a number of large ceramic capsules sixty-five feet deep in Antarctic ice, where the temperature is −60° C. In the capsules would be "seeds, spores, human and other reproductive cells, human mother's milk, DNA, rainwater, sea water, air, and soil" kept perfectly intact for analysis by scientists in centuries to come. Once the Antarctic cache is established, the team would like to place capsules on the Moon, where the temperature is −230° C. and there is neither air nor moisture to foster rot.

(Come to think of it, if human beings do become a spacefaring species, Earth's Moon might be an ideal eventual location for the 10,000-Year Library. Over that time frame humanity's main story would be of global convergence followed by a massive diaspora into space. The diaspora's point of origin would be a prime candidate for record keeping, and the Moon offers a stable, durable site, easily accessible from space, with a good view of grandmother Earth.)

Even in the short term the Library could provide a number of time-release services. Secrets that are meant to be kept for a certain time, or until certain people have died, could be held physically or crytographically sequestered until their time is ripe. Property deeds, contracts, wills, directions to caches could be securely stashed with appropriate wake-up directions built in. Danny Hillis points out that "Search for Extra-Terrestrial Intelligence programs need this. By the time any sort of extra-terrestrial life is likely to answer, we will have forgotten what we asked."

The Library could offer personalized I-Told-You-So! services. Register your prediction, your hunch, your wild scheme, your strong argument about the future course of events, set the wake-up date, pay the fee, and relax, knowing that history will be given irrefutable notice of how right you were. A sufficient mass of such material over time could give researchers data for insight into the nature of future telling and its progress, if any, over time.

The same dynamic is behind the idea of a Responsibility Record. Suppose we wanted to improve the quality of decisions that have long-term consequences. What would make decision makers feel accountable to posterity as well as to their present constituents? What would shift the terms of debate from the immediate consequences to the delayed consequences, where the real impact is? It might help to have the debate put on the record in a way that invites serious review.

One side of such a debate could file (for a fee) its arguments, facts, media reports, major players, and predictions with the Library's Responsibility Record, along with desired times in the future for the record to "wake up" for review. The Library would then of course contact the opposition to see if they would like to do the same. The record is public; each side could attach corrections and rebuttals to the other side's file.

Years later, at the wake-up times, those living with the consequences of the decision will be able to see what both sides registered, will compare the two versions with what actually happened in the world, will assign blame and credit accordingly, and will by the way notice whether the terms of the original debate had anything whatever to do with what actually happened. This is where the real payoff lies. By harnessing the power of contention, the Library can accumulate detailed records of countless sequences of debate-decision-consequence, on a growing range of subjects, spanning ever-longer periods of time.

A well-managed Responsibility Record would be both trove and warning. The warning is to policy makers and issue debaters that they will be held accountable by posterity. The trove is for delvers in lessons to ponder and explore ever-better terms of debate for issues with long-term consequences, and to frame current debates in the context of relevant past debates. We can stop reinventing the square wheel.

The Responsibility Record fosters slow, direct-feedback loops in policy. This is a new and perhaps crucial service, because up to now civilization's feedback loops have been either direct but quick (win the election), or slow but indirect (suffer a gradually degraded environment). The urgent always had a louder, clearer voice than the background rumble of the important. The Responsibility Record doesn't try to change the voices; all it does is retune our hearing.

Another long-view service has been suggested by Esther Dyson. Every so often the public is gripped by a great mystery. Who kidnaped the Lindbergh baby? Who was behind the killing of President Kennedy? Did Dr. Sam Sheppard murder his wife? Who was the Unabomber? Time has solved two of these. Math instructor Ted Kaczynski was the Unabomber; and the convicted Dr. Sheppard was innocent back in 01954.* DNA analysis forty-four years later proved that a window washer, Richard Eberling, not only killed Mrs. Sheppard but raped her first (something never mentioned in the celebrated trial). The Long Now Library could preserve rich archives of such mysteries so that they can be relived in

---

*Just to try out the 10,000-year perspective, the remainder of this book employs the five-figure year dates proposed in the previous chapter.

light of what eventually is discovered. As new mysteries emerge, people can expect the Library to stay on the case. They will be able to experience their contemporary mysteries simultaneously in the present and in terms of how they might look in the future, when they have been solved. The ability to live both in the present and in a handful of imagined but uncertain futures is the basic skill of foresight, planning, and responsibility. It is worth encouraging.

Just as new information can be applied to old situations, uninvited but important old information might be applied to new situations. A few centuries from now the Library might send a message: *To the government of the region formerly known as New Mexico. Upon periodic review of the long-term hazards file for your area, it has been determined that a very large—an extremely large—quantity of radioactive waste was buried in salt structures a short distance east of what used to be called Carlsbad Caverns. In the event that this comes as news to you, we can furnish the exact location of this hazard, which is expected to be potentially harmful for another nine thousand years. Your next notification will be delivered three hundred years from now.*

Better still might be messages in time bottles. Hide forest-losing statistics and reforestry advice inside virgin forests. Bury information on global warming inside glaciers. Seed uncleared minefields with data on who exactly planted the mines.

The Library should specialize in trends too slow to notice but that gradually dominate everything as they accumulate: these are the genuine megatrends in economics, demographics, and environmental data. Our civilization is skilled at focusing on content, Esther Dyson points out, but we have not developed good peripheral vision for gradual shifts in context. The Library should take an active role in supporting extremely long-term scientific studies (the subject of an upcoming chapter).

One of the best ideas I have heard for a library of the future is not a library at all, but a museum. Let Richard Benson, dean of the Yale School of Art, make the case:

> Embed the Clock, as a centerpiece, in a new museum of the history of technology. If technology is to be the future of the living world, then we have to admit that it is at its starting block. We are at the Cambrian explosion of technology, and we are at the perfect point in time to gather the fossils as they

are being made and discarded. The point at which technology really took off is with the invention of the heat engine, and the bits and pieces of this brief period are still around to be preserved. Engines, locomotives, cars, planes and all the pieces of the great transportation revolution are also still around. The brand new electronic revolution is taking place in our midst, and we could easily gather up the detritus of this great step's beginning. There are unthinkable things ready to happen, and they will occur at a dizzying pace; we could build an institution around the recording of these changes. The very nature of the institution could be to persist over time so the record is made as complete as possible. We also are living in a fat time, with great wealth and stability, and it would seem reasonable that those making fortunes with technology would be interested in preserving a record of their achievements. Without institutional and financial long-term support, the Clock will disappear as quickly as the small group who are trying to make it.

Such a museum could, if necessary, be not just a collection of curiosities but a template for renewal. When civilizations truly crash, no one at the time can imagine the depth of the fall, nor how labored and long the revival, if it ever happens. After Rome fell, "Europe went unwashed for a thousand years." Cities emptied; literacy vanished. All the Roman achievements of engineering, culture, and government simply ceased to be. Population in some areas dropped by nine-tenths. Even with the heroic continuity of the Catholic Church, the skein of culture was reduced to fragile wisps: Only one copy of Lucretius made it through the Dark Ages, only one copy of five books of Livy, one copy of nine plays of Euripides, one copy of Tacitus, one copy of *Beowulf.*

Of the fifteen to twenty thousand distinct languages once spoken two-thirds are extinct, and the pace of loss is increasing. "The death of a language," writes George Steiner, "be it whispered by the merest handful on some parcel of condemned ground, is the death of a world."

Civilization now is global. It is ever more tightly linked and ever more leveraged out over the abyss on an elaborate superstructure of highly sophisticated technology, every part of which depends on the success of every other part. All this may make it more robust against catastrophe, or more frail, we don't know yet. What we do know is that a global collapse cannot count on some other civilization coming to the rescue, since by then there will be no other. It is strange to contemplate, except in light of thousands of years and the demise of twenty-some previous civilizations.

Perhaps the 10,000-Year Library should be thought of as an insurance provider. It offers detailed risk assessment of the chances civilization is taking (which might be enlightening in its own right) and promises resources for recovery if, God forbid (as insurance agents say), the worst should happen.

The scientist James Lovelock, best known for his Gaia theory of life-mediated regulation of the atmosphere, has proposed compiling a start-up manual for civilization, beginning with how to make fire, moving on through all of science and technology, from subjects such as ancient genetic design (domesticating plants and animals by selective breeding) to current genetic design (cloning). "Who would guard such a book?" Lovelock asks. "A book of science written with authority and as splendid a read as Tyndale's Bible might need no guardians. It would earn the respect needed to ensure it a place in every home, school, library, and place of worship. It would then be on hand whatever happened."

Lovelock worries about science skills being lost because they have become so widely scattered into countless narrow specialties. His civilization primer would be the great cross-disciplinary reference work. Doug Carlston lists other categories of endangered information: "information that was important to many but held by few, old information (Dead Sea Scrolls), restricted information (Stasi files), information of importance over long periods of time (gene structures, seismic records, weather, seeds!), information held largely in highly degradable form (Technicolor movies)."

What do historians want preserved? I asked William McNeill, author of *The Rise of the West* and *Plagues and Peoples*, about the kinds of things his fellow historians wish had been saved from the past. Well, he said, there was the census of the entire Roman Empire, which the parents of Jesus were avoiding. Caesar delivered the document to the Roman Senate, and that's all we know of its priceless contents. Then, there are delicate points in history, like when Alexander the Great failed in India and headed back. Some historians think that he was planning to conquer North Africa, which would have made the Mediterranean an Alexandrian lake and changed history. Some personal diaries of his generals would be helpful to have.

The problem everyone has is that *you never know* what will be treasured later. When we look at old magazines, the ads are far more fascinating and informative than the articles. The U.S. Weather Service receives considerable income from selling old weather reports. To whom? To lawyers, who want to know if it was raining on the night in question. The BBC, on a housecleaning binge a few years back, tossed out some of its video archives considered trivial and has been gnashing its teeth over that cultural loss ever since.

The largest and heaviest book in the world is inscribed on 14,300 large stone tablets concealed in caves near the Yunju monastery in Beijing Province, China. In a time of book burnings, 00605 C.E., a monk set about preserving the Buddhist scriptures on stone. The work continued for a thousand years, and then the entire trove was hidden in 01644 C.E. The hoard is indeed valuable for studying the history of Buddhist thought, but probably we would value the stones more if the monks had simply recorded the weather and what they were eating. Better still would have been a reverently preserved sequential archive of dried monk poop, which would yield no end of data on diet, agriculture, climate, health, and racial and family lineage. You never know what people will want preserved.

One of the great instruments of civilization is the idea of the canon: the select set of items deemed to represent the best of a genre and the main line of progress and transmission from generation to generation in that genre. A primary function of universities is the care and feeding of various canons—mainly literature, the arts, science, and the named academic disciplines. Other canons are strangely untended, such as technology, agriculture, business, and such nonacademic pursuits as gardening, furniture design, currencies, and pets.

One canon I would like to see established is that of the great textbooks. Just knowing the current list—*The Cell* in microbiology, *The Art of Computer Programming*, Renfrew's *Archaeology*—would enable anyone to pursue top-level education on their own. All the best textbooks in combination would nearly add up to Lovelock's primer of civilization. Study of the evolution of the most influential textbooks over time would yield peerless insight into intellectual history. Comparative analysis of what makes the best contemporary textbooks so good might lead to even better textbooks being written.

Do such ideas justify setting in motion the building of a 10,000-Year Library? By themselves, perhaps not. The ultimate reason for initiating something ambitious is not to fulfill certain notions but to find out what surprises might emerge. The most remarkable results almost certainly cannot be anticipated. What would the existence of something thought of as a 10,000-Year Library bring into the world? "Boiling rocks" is what the novelist and provocateur Ken Kesey calls this kind of research. "If you don't boil rocks and drink the water, how do you know it won't make you drunk?"

TRAGIC
OPTIMISM

Some ideas present themselves to a writer as if they were for a live audience, in the tone and cadence of a speech, even anticipating imaginary questions. This happened to me here. Apparently I needed a break from the essay voice and its narrow form of logic, and perhaps you wouldn't mind one as well. So please imagine you are in a chair turned away from a conference banquet table to attend the evening's entertainment.

Thank you for that very kind introduction. I hope you won't mind if I don't talk about what the program says I will tonight, but about something more interesting.

On her recent album, *Bright Red*, Laurie Anderson posed the big one—*What I really want to know is this: Are things getting better or are they getting worse?*

What do you think? Are things getting better or are they getting worse? Query yourself three times; the question is worth it. First answers usually are knee-jerk. Second answers tend to be cute. Third answers to the same question sometimes tell the truth.

While you're working through your answers, I'll tell about Herman Kahn and free will. The late, great futurist Kahn used to ponder the question of free will with his audiences. "It's a fundamental question," he would say. "Do we have free will, or is everything determined? I don't have an answer I'm sure of, but I am convinced that people behave better when they think they have free will. They take responsibility more and they think about their choices more. So I believe in free will," said Herman Kahn.

Okay, question time. Are things getting better, or worse? How many hands for better? Uh huh. How many for worse? Interesting. About half and half. This is definitely a technology audience, feeling chipper about things. I assure you that this same audience in the late 1980s would have voted 75 to 90 percent pessimistic, and may well again one day.

Most people these days think things are getting worse. At Global Business Network we work with strategists in large organizations all over the world, and their view of the future is routinely bleak. We study opinion surveys from around the world. Same thing: People everywhere are worried about the future. There's been a recent upturn of optimism in some sectors, such as in this room, but it's clearly shallow and fragile.

Maybe that's as it should be. There is lots to worry about. If people fret enough, maybe they'll take measures to fix things before they get worse. Preserve us from witless optimists!

On the other hand, how does the question play against Herman Kahn's pragmatism test? Do people behave better when they think things are getting better or when they think things are getting worse? If you really think things are getting worse, won't you grab everything you can while you can? Reap now, sow nothing? But if you think things are getting better, you invest in the future. Sow now, reap later.

How you think about the future depends in part on how you think about time. Laurie Anderson has another question on her album: *Is time long or is it wide?*

I'd like to know her interpretation of that question. Mine is that time can be thought of in terms of everything-happening-now-and-last-week-and-next-week (wide) or as a deep-flowing process in which centuries are minor events (long). The wide view sees events as most influenced by what is happening at the moment. The long view perceives events as most influenced by history: "Much was decided before you were born." The wide view is disparaged as short-term thinking. The long view is praised as responsible.

Wide time is on the increase these days, and for good reasons. Technology seems to be accelerating, and you have to keep up. High-turnover networks and markets rule, instead of staid old hierarchies, and you have to keep up. High-tech people like us, I warrant, are largely wide timers. When you look at our sense of the deep past and deep future, both are faux: medieval fantasies in one direction, space fantasies in the other. We are interested in, for instance, the exponential growth of the World Wide Web. It's useless to try to imagine what the Web might be like in, say, 2045 (only as far removed in time as the bombing of Hiroshima), so we don't bother. Does this mean that technoids and their camp followers are responsibility-impaired? Could be.

Environmentalists are supposed to be the long-view specialists these days, but I think we do it poorly. I was trained as an ecologist, so I know how extremely limited our longitudinal studies are: about the length of time it takes to get a graduate degree. Since it is the long, slow fluctuations and cycles that most influence everything in ecology, we still don't have the most important information on how natural systems actually work over time.

Also, environmentalists are calamity callers. We're the leading apostles of Things Are Getting Worse. Gregg Easterbrook has written a whole fat book of environmental good news called *A Moment on the Earth*, in which he fricassees his fellow environmentalists for scanting their many successes and occasionally lying about the problems (spotted owls abound in second-growth forests, for example). Some years back the pioneer environmentalist René Dubos—who coined the phrase "Think globally, act locally"—wrote a paean to the places where humans and natural systems have blended beautifully for generations. It was titled *The Wooing of Earth*. It is long out of print.

Firmly keeping the facts in mind and balance in mind, Dubos and Easterbrook do have the responsible approach. Things are getting better. Invest in it.

As for Laurie Anderson, in her recent concerts she told of interviewing the avant-garde composer John Cage when he was eighty—"an age when most people are in a bad mood." She put the better-or-worse question to him. Cage hedged cheerfully for a while and then admitted he thought things are getting better—*slowwwly*. That's just right.

Are there any questions?

VOICE: What does Laurie Anderson really think?

Heh. When I was lucky enough to have the opportunity, I had to ask her three times. She was even less forthcoming than John Cage, but she did eventually confess. Better.

VOICE: What do you think, really?

In public, except for tonight, I usually say things are getting worse unless some particular change is made. Warnings are important. In private I usually think things are getting better, and that makes me invest my time in interesting projects, but then nobody close to me has died lately.

VOICE: So why are you promoting optimism tonight?

Cowardice. This audience is comfortable with it. If I gave this talk anywhere in Europe—Sweden, for example—I would be attacked viciously for my ignorance, naiveté, and callow irresponsibility. I suppose I could defend myself with Arthur Herman's wonderful book, *The Idea of Decline in Western History*. He says that in Europe high-minded cultural pessimism began with the failure of the French Revolution and culminated in Nazi Germany. It was tremendously destructive. It still is.

VOICE: But the pessimists are always right! (*Applause*)

Yeah, and they love it. My worry is they love it so much they avoid any efforts that might make them wrong.

Look, it all depends on what time frame you look at. Paul Saffo says that in the short term the pessimists are right, and in the long term the optimists are right. Everything has been going to hell as long as anyone can remember. Empires are always dying. Your friends are always dying. But in the long sweep of history, on average, life has been getting steadily better for as long as you care to look. Does anyone here really want to live in medieval times? Have rotten teeth, eat turnips, and die at the age of twenty-seven of exhaustion?

So, short term worse, long term better. Maybe the way to resolve them is tragic optimism. I would settle for a world of tragic optimists.

VOICE: Isn't any kind of optimism just a convenient monopoly of the privileged?

I don't know. The comedian Bill Cosby says that in college he ran into the question, "Is the glass half empty, or half full?" He took the question home, and his father told him, "It depends if you're pouring or if you're drinking."

FUTURISMO

The greatest futurist of the twentieth century won't let anyone refer to him as a futurist. Peter Drucker looks forward the way a skilled historian fifty years hence would look back on our time. All of his books about the future are still in print: *The Future of Industrial Man* (01941), *The New Society* (01950), *The Landmarks of Tomorrow* (01959), *The Age of Discontinuity* (01969), *The New Realities* (01989). So what is Drucker's problem with being called a futurist? Is it a tainted calling?

The history of soothsayers is long and comic. Recall China's first great emperor, Shih Huang-ti in the third century B.C.E., who was so obsessed with geomancy that he preserved fortune-telling books while burning all the classics. In the West there is the Greek and then Roman tradition of the sibyls: crones who supplied written prophecies at the great oracle centers. A set of these prophecies, the *Sibylline Books*, was offered for sale by the Cumaean sibyl to Tarquinius Superbus, a king of Rome. "He refused to pay her price," one history reports, "so the sybil burned six of the books before finally selling him the remaining three at the price she had originally asked for all nine."

These days there is a futurist profession of sorts, but it is larded with hobbyists and special pleaders; their flakiness is what makes Drucker shun their conferences. I think it is time to draw a harsh distinction, similar to that drawn between science and "scientism" (the style of science without its substance). There is a domain of future studies, rigorous and objective, and another that is essentially "futurism"—a belief structure, often highly subjective. There are futurists, like Drucker, and those who pretend to be futurists: *futurismists*, exuding *futurismo*.

Futurismists are not bad people, and certainly not fraudulent. If anything they are captive of their goodness. High-minded and earnest, they have meetings to determine "the goals of humankind," and to advance worthy causes such as feminism, multiculturalism, and a world free from hunger. The distinguishing trait of futurismists is that they have an agenda: something they want to have happen or something they want to prevent from happening in the future, often based on a particular ideology, political bent, theory of history, or special interest. Some hive off into sectlike groups, such as the Extropians—a 01990s California enclave of bright and enthusiastic Singularity advocates who could hardly wait for the techno-Rapture. They have a classic case of what Paul Saffo calls *macro-myopia*: "we overexpect dramatic developments early, and underexpect them in the longer term."

Futurismists are predictable once you know their agenda, whereas futurists are not. Herman Kahn, inventor of scenario planning in the 01960s, was a conservative and proud of it, yet he could always startle an audience with a new perspective. Two Long Now board members are well-regarded professional futurists: Peter Schwartz of Global Business Network and Paul Saffo of the Institute for the Future. Schwartz is a liberal, and I believe Saffo is; both are valued because they are full of surprises. The surprises come from their ruthless curiosity about the world and what is truly going on in it, not from their politics. "There are two sets of futures," Desmond Bernal wrote in 01929, "the future of desire and the future of fate, and man's reason has never learned to separate them." In action they may be intertwined, but distinguishing one from the other is easy. The future of desire is always predictable; the future of fate seldom is.

Futurismists displace themselves forward in time. One of their appealing-sounding truisms says, "We all have an interest in the future. It is where we will spend the rest of our lives." No, we won't. Like it or not, we will spend the rest of our lives in the present, as it unfolds from day to day. Good foresight, good planning, and good luck (as Herman Kahn always included) can sometimes improve the quality of the present, but the future itself remains forever out of reach.

Some futurismists have a utopian agenda: a desire to craft a more rational world. In the sixties and seventies in America my generation had some experience with utopian agendas and their results. We had our noses rubbed in our fondest fantasies. Like all experienced utopians, eventually we abandoned utopia and became good-natured cynics. (Good-natured because we at least tried. Ideologues who never try to live up to their fantasies become bad-tempered cynics.) We watched in early admiration and increasing horror what happened in Mao's China and Pol Pot's Cambodia, where resetting the clock of history to Year Zero in 01975 began a three-year terror by the Khmer Rouge that killed one million Cambodians. What we learned in our own communes was a mild form of what so many Asians learned: All utopias become dystopias.

We never stopped reading futurismic science fiction, though, and its value keeps growing with every decade. As mathematician and cultural essayist Freeman Dyson observes, "Economic forecasting makes predictions by extrapolating curves of growth from the past into the future. Science fiction makes a wild guess and leaves the judgment of its plausibility to the reader. . . . For the future beyond ten years ahead, science fiction is a more useful guide than forecasting." A good science-fiction story is a scenario in depth—a whole possible future. Subjectivity and agendas are advantages for science-fiction writers because they lend power to the stories. Thus right-wing libertarian Robert Heinlein got generations of male adolescents interested in the future by harnessing their urge to defy authority. Thus left-oriented Olaf Stapledon (*Starmaker*, 01937), Ursula LeGuin (*The Dispossessed*, 01974), and Kim Stanley Robinson (*Mars* series, 01993–97) create gripping worlds with their utopian attention to thoroughly worked-out detail.

According to Kevin Kelly, "Isaac Asimov once said that science fiction was born when it became evident that our world was changing within our lifetimes, and therefore thinking about the future became a matter of individual survival." The nanotechnology futurist Eric Drexler concurs: "I have found over the years that people familiar with the science fiction classics find it much easier to think about the future, coming technologies, political effects of those technologies, and so on." At Global Business Network (GBN), the scenario-planning business that employs me, we frequently send out science-fiction books to the Network membership, and when we can get writers such as William Gibson, Bruce Sterling, David Brin, and Vernor Vinge to attend scenario workshops with client organizations, the quality of work deepens. Scenario planning involves exploring several widely variant futures of an organization's world persuasively in depth. Skill in science fiction adds to the depth.

Still, the most important developments in the future, says Freeman Dyson, keep being missed by both the forecasters and the storytellers: "Economic forecasting misses the real future because it has too short a range; fiction misses the future because it has too little imagination."

Too *little* imagination? Yes, for a structural reason almost never taken into account. At any time the several "probable" things that might occur in the future are vastly outnumbered by the countless near-impossible eventualities, which are so many and individually so unlikely that it is not worth the effort of futurists or futurismists to examine and prepare for even a fraction of them. Yet one of those innumerable near-impossibilities is what is most likely to occur. Reality is thus statistically forced always to be extraordinary. Fiction is not allowed that freedom. Fiction has to be plausible; reality doesn't.

In that light, a prediction machine described by Oliver Sparrow in a GBN online discussion is formally correct:

> In the town of Surkhet in the province of Achar in Eastern Nepal, there stands a concrete space rocket, with a minaret mounted upon it and donkeys grazing around its tailfins. Inside, a moribund PC/XT computer is hitched to a stereo system. As the hour approaches, the random-number generator in the computer produces a string of sounds—dong, ding, hoot, peep—which can run for up to a minute, signifying nothing but the passage of time. Amongst the donkeys below, however, and between the orange vendors, sellers of horoscopes have set up in business, predicting your personal future from the weird cacophony. Two poops and you're out.

Entrails, tea leaves, tarot cards, arbitrarily generated sounds like at Surkhet—their randomness and ambiguity accurately represent *how* the future unfolds but not *what* will happen. The soothsayers had the irrational mechanism right, but their overrational and subjective mode of interpretation is wrong; they predict only in terms of what is feared or desired. The core fallacy of futurismo is: Desire always misreads fate.

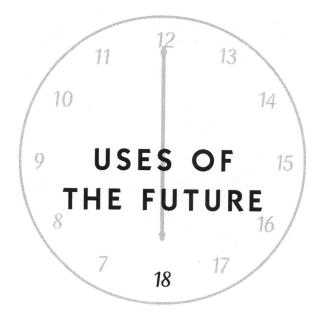

USES OF
THE FUTURE

Precisely because the future is essentially unknowable, it forces us to stay broad-minded and fast on our feet. And this is only one of its many services.

Imaginative scenario planning, by giving up on any hope of accurately predicting the future, yields strategies made robust by their wide scope of alertness and swift adaptivity. You don't plan for a single certain future but rather for multiple possible futures, each based on a different theory of what's really going on. "Global markets lead to global prosperity"; "Global markets favor some regions and crush others"; "Global markets are inherently unstable." "What can our government (or company) do now to thrive in any of those worlds, given that we can't know which one might play out in reality?"

One surprising by-product of the scenario-planning process is increased responsibility. Corporations discover the need to take care of their industry as a whole, or to protect the natural environment, or to promote civil liberties. This comes not from virtue but solely from the ability to engage longer periods of time. While twenty-year forecasts are a complete waste of effort, twenty-year scenarios are common and useful. Any organization confidently thinking twenty years ahead is compelled to grapple with long-term needs, such as an educated workforce and a sustainable regional economy. Rigorous long-view thinking makes responsibility taking inevitable because it responds to the slower, deeper feedback loops of the whole society and the natural world.

Another surprising feature of the scenario approach is that parties in dispute can utilize it together. Using multiple scenarios allows parties to continue to disagree about the past and present (since each scenario can represent a different version of what happened in the past) and at the same time allows them to agree about possible futures they face together, and which of these might be most desirable for all. Multiparty scenarios of this sort helped end apartheid in South Africa. The shift from the de Klerk to the Mandela government was greatly eased by what became known as the Mont Fleur Scenarios, in which representatives from all the contesting groups participated. The scenarios helped move discussion and action away from dispute about conflicting interests and toward building on common ground—the future all realized they would have to share.

The bodies of most animals are configured toward the future, our faces leading the way in the direction of travel. Our mental framework has what the philosopher Derek Parfit describes as a "bias toward the future." Future pain, such as a forthcoming visit to the dentist, gets more of our attention than past pain. We may put off a future pleasure temporarily just to savor it longer, whereas the same pleasure in the past is less interesting. "We go from anticipation to anticipation," said Samuel Johnson, "not from satisfaction to satisfaction."

Time is asymmetrical for us. We can see the past but not influence it. We can influence the future but not see it. Both the invisibility and potential malleability of the future draw us to lean into it, alert to threat or opportunity, empowered by the blankness of its page (if the future is not determined, we might do anything). This is the forward-leaning Now that would be interesting to expand from next week to next millennium. The invisibility is vastly greater at millennial scale, but so is the empowerment, since things we do now might have enormous impact over the next thousand years.

Wait. Is it greater empowerment over the long span of time, or less? I would argue it is greater, but economists insist that value over future time goes down. The loss of value can even be measured: It is the discount rate (also called interest rate) on money, typically 10 to 15 percent a year, based on the perceived uncertainty of what may happen in the future. You are somewhat uncertain about what will happen next year, extremely uncertain about what will happen in twenty years, and infinitely uncertain about events in one hundred years. If you loan someone $10,000 to be paid back over ten years, and the uncertainty makes you charge a normal 15 percent interest, you will get back over $25,000 total over the ten years. That ten-year $25,000 was discounted to you at 15 percent a year, so you only paid $10,000 for it.

Does the passage of future time dilute value or increase value? The question becomes a crucial issue when, for example, a corporation buys large tracts of redwood forest that have been logged sustainably for decades and proceeds to cut all the trees down at once because the forests are not yielding as much revenue as the same amount of money would gathering interest in a bank. This was done in California in the eighties and nineties by a corporation called Maxxam. Discounting the future led to modest short-term individual gain and horrendous long-term public loss. The accounting was too isolated. Commercial pace was allowed to rush a natural and cultural asset to destruction.

A futurist at SRI International, a California think tank, once sculpted a "future tree." It had a fat trunk (the present), several branches indicating various major directions in which the future might go, and thirty-six twigs at the top, showing that many distinct futures in a few years. No wonder we discount the future: This little twig might be the future? It's nothing.

The tree's dimensions should have been inverted. The present should be a thin trunk, then fatter branches, and enormous twigs. Future considerations should dwarf the present—the same way unborn humans vastly outnumber the living, the same way the accumulative never-born of an endangered species should loom over the debate about its protection.

Virtual realist Jaron Lanier refers to this line of thinking as *karma vertigo*. "The computer code we are offhandedly writing today could become the deeply embedded standards for centuries to come. Any programmer or system designer who takes that realization on and feels the full karmic burden, gets vertigo."

The karmic view of the future can be as distorting as the discounted view. Instead of the reduced responsibility of discounting, karma can impose crushing responsibility, paralyzing to contemplate. Male mammals produce billions of spermatozoa. If we do not help every one of them fertilize an egg and become an adult, how can we stand the karmic inferno of all those unborn beings howling at us? No one would dare do anything, for all the terrible things it might cause and all the beautiful things it might prevent.

Is there a resolution to the paradox between karma and discounting? There is at least relief from it in the pace layering of civilization (where the pace of change slows from rapid Fashion down through Commerce, Infrastructure, Governance, and Culture, to glacially slow Nature). In the fashion and commercial domains a discounted approach to the future is necessary to maintain the customary swift turnover. An increasingly karmic and careful approach, however, is appropriate to managing the slower layers of infrastructure, governance, culture, and nature. It would be nice to have one body of economics that embraces all the levels, but we don't yet.

Also, a 10,000-Year Clock offers token relief. In its perspective each future year is neither lesser in import than the present year nor greater, but exactly the same. Karma and discounting both are voided by the Clock, or perhaps balanced in its presence.

Another version of the discounting debate emerges around long-term planning. Many have noticed, these decades, that there seem to be fewer long-duration projects, even though there is growing wealth to invest in such work. Kevin Kelly once raised the question at a dinner with complexity scientists. In friendly but acerbic terms they mocked the ambitions of The Long Now Foundation. Kelly paraphrased their argument:

> Since complexity theory shows that even the fairly near future is inherently unpredictable, any polygenerational plan will guess wrong about what a future generation wants or needs. Suppose a previous generation had expended great effort planning for dirigible ports around the world in the year 2000! Inevitable technology obsolescence and economic discounting renders any long-term return of value impractical. Conservation makes sense for the long term, and so does science (because it is incremental and open-ended), but specific long-term plans will always be based on wrong long-term predictions, and it is best to avoid them.

Danny Hillis responded to Kelly's report:

> The difference between the two examples—dirigible ports versus ecological conservation—is a great demonstration of the difference between long-term planning and long-term responsibility. I agree that the former is futile, but that's no excuse to give up on the latter. The difference is between trying to control the future and trying to give it the tools to help itself. Believing in the future is not the same as believ-

ing you can predict it or determine it. The Long Now Foundation is not about determining the destiny of our descendants, it is about leaving them with a chance to determine a destiny of their own.

Such debates indicate that the way the future is viewed and used is in transition. Some say that a sense of any future at all was extinguished for three generations in the twentieth century by the dread of nuclear armageddon, from which we have not yet recovered. At the same time, increasing reports of incremental loss—of atmospheric ozone, of species diversity, of rural village stability—tell us that long-term maintenance issues are accumulating to crisis proportions that short-term thinking is powerless to address. "For most of civilization's history," observes Kelly, "tomorrow was going to be no different than today, so the future was owed nothing. Suddenly, in the technological age, our power of disruption became so great, there was no guarantee we'd have any future whatsoever. We now know we are stuck with having a future, and thus are obliged to it, but we have no idea what that means."

Some of what the future means can be revived from traditional ethics, such as Samuel Johnson's admonition, "The future is purchased by the present. It is not possible to secure distant or permanent happiness but by the forbearance of some immediate gratification." Some we can learn from the emerging field of future studies. "The first thing you learn in forecasting," says Paul Saffo, "is the longer view you take, the more is in your self-interest. Seemingly altruistic acts are not altruistic if you take a long enough view." In the long run saving yourself requires saving the whole world.

Governance itself is being rethought. "The proper role of government in capitalistic societies in an era of man-made brain power industries," writes the economist Lester Thurow, "is to represent the interest of the future to the present." Commerce has too short a time horizon to take the larger future seriously, therefore governance must do it. Governance can forcefully represent future Californians who might want a thriving redwood forest instead of Maxxam having once maximized its profits by clear-cutting the forest.

A major point of reference for thinkers about the future is Robert Axelrod's 01984 book, *The Evolution of Cooperation*. It reported his seminal research on the playing of a simple game called

Prisoner's Dilemma, in which neither of the two players can know what move the other will make next. The game is diabolically structured so that if both act in mutual trust, they are moderately rewarded; if both defect from trust, they are mildly punished; and if one defects and the other doesn't, the defector is richly rewarded, and the trusting one strongly punished. The apparently safest way to play the game is to always defect, but if both players do that, neither of them does as well as they would if they both cooperated. Hence the dilemma.

Axelrod proved that if the game continues over time—what is called *iterated* Prisoner's Dilemma—a strategy called *tit for tat* emerges spontaneously, allowing both players to cooperate and thus get higher scores. The game automatically generates cooperation if what Axelrod calls "the shadow of the future" is allowed to lengthen. By continuing to play each other, each player develops a reputation that the other player learns to count on and work with. Even in a game that rewards distrust, time teaches the players the value of cooperation, however guarded they may be.

To produce the benefits of more cooperation in the world, Axelrod proves, all you need to do is lengthen the shadow of the future—that is, ensure more durable relationships. Thus marriage is common to every society, because trusting partners have an advantage over lone wolves. Thus Allied and German troop units in World War I who faced each other across No Man's Land for months quietly developed a set of local live-and-let-live arrangements. Stuck with each other, they made a separate peace.

The great use of a continuous future, then, is its inclusiveness. Given uncertainty, we are right to use many scenarios (hence discounting any one of them), but we are also right to assume we share one world (hence karmic responsibility). Anything may happen in the future; reliable pattern only emerges in how people handle events over time. Steadily engaging the future teaches us wariness about events and trust in each other. We don't know what's coming. We do know we're in it together.

USES OF THE PAST

Like a tree, civilization stands on its past. Only 5 percent of a mature tree's mass is alive—the leaves, cambium, sapwood, and root tips. All the rest is dead, yet that gradually built structure of once-living wood is what allows the leaves to reach so high and the roots to draw so deep. The medieval historian George Holmes describes thousand-year continuity in modern Europe:

> Most Europeans live in towns and villages which existed in the lifetime of St. Thomas Aquinas, many of them in the shadow of churches already built in the thirteenth century. . . . The modern nation state grew out of the monarchies created by kings such as Philip Augustus of France and John of England. . . . Our methods of commerce and banking are derived from the practices of the Florentine Peruzzi and Medici. Students work for degrees already awarded in the medieval universities of Paris and Oxford. . . . Our books of history and our novels are the linear descendants of the works of Leonardi Bruni and Giovanni Boccaccio.

Some of the deepest connections we can make are with our own distant past. This is one of the great benefits of long-lasting religions, but even a secular connection with the past can be deeply moving. The thrill Brian Eno and I felt at the top of Big Ben was surpassed the next day on a tour of the back rooms of the Egyptian collection at the British Museum. At one point our host reached into one of the countless crowded shelves and drew out a desiccated human forearm and hand. Eno wrote later:

> The nails were beautifully manicured and hennaed, and in perfect condition. The hand, which was quite black, was gently curved as if holding a small animal or bird. I held this hand, slipped mine into it, and felt momentarily a connection with this very ancient woman. The fact that there was

only a hand and forearm, not a whole skeleton, gave the sense of a living woman—one somehow imagined the rest of her being. It was a strangely intimate experience: I was holding the hand of a person from so many years ago, a person who'd lived in this sophisticated and complicated society which loved pets and drank beer and went hunting with boomerangs and played games with bats and balls.

At Laetoli, Tanzania, the archaeologist Mary Leakey discovered an impossibly ancient trail of hominid footprints left in solidified volcanic ash by a male, female, and child. She reported:

At one point, and you need not be an expert tracker to discern this, she stops, pauses, turns to the left to glance at some possible threat or irregularity, and then continues to the north. This motion, so intensely human, transcends time. Three million six hundred thousand years ago, a remote ancestor—just as you or I—experienced a moment of doubt.

We draw profound comfort from such experiences. In the light of the long human story our own concerns seem petty and local. Something more compelling than nostalgia connects us to continuity and durable traditions. "I love everything that's old," chants Oliver Goldsmith, "old friends, old times, old manners, old books, old wines." We look back on the known and draw sustenance.

Yet comfort is not all we see looking back. At many times and places the past was considered horrible and dangerous. In 01940 the Marxist historian Walter Benjamin described his vision of the angel of history:

His face is turned toward the past. Where we perceive a chain of events, he sees one single catastrophe which keeps piling wreckage upon wreckage and hurls it in front of his feet. The angel would like to stay, awaken the dead, and make whole what has been smashed. But a storm is blowing from Paradise; it has got caught in his wings with such violence that the angel can no longer close them. This storm irresistibly propels him into the future to which his back is turned, while the pile of debris before him grows skyward. This storm is what we call progress.

The idea of history as horror and warning itself has a distinguished history. The philosopher George Santayana voiced the sharpest version of the perennial warning in 01905: "Those who cannot remember the past are condemned to repeat it." Blinkered willfulness leading to calamity is so common in human experience that we can count on it recurring endlessly unless attention is paid and lessons harshly drawn, diligently remembered, and then applied.

It is no accident that among the finest leaders of the 20th century is a professional historian. Describing Winston Churchill, Isaiah Berlin wrote, "the single, central, organizing principle of his moral and intellectual universe is a historical imagination so strong, so comprehensive, as to encase the whole of the present and the whole of the future in a framework of a rich and multicolored past." Reading history, writing history, and creating history were all one enterprise for Churchill.

The idea of applied history, such as Churchill employed, is in low repute among historians, for the same reason that future-thinkers with agendas are suspect among futurists. The interested eye is easily blind to what does not suit its interests. Nevertheless, policy keeps being made, and the interested eyes making these decisions require historical perspective if they want to avoid the Santayana curse.

So far as I know there is only one good text these days on how to apply history intelligently: *Thinking in Time: The Uses of History for Decision Makers* (01986) by Richard Neustadt and Ernest May, who have been teaching a course on the subject at Harvard's Kennedy School of Government since the early 01970s. Both the course and book have been popular and influential, though they have not yet generated an academic discipline of Applied History.

The need is well catalogued by Neustadt and May:

General knowledge of history is less and less characteristic of American decision-makers and their aides. Our educational system turns out lawyers who may know only the history they learn through the constricting prisms of court opinions; economists who may learn neither economic history nor much if any economic thought except their own; scientists who may know next to nothing of the history of science; engineers who may be innocent of history entirely, even that of their profession; graduates of business schools with but a smattering of theirs; and generalist B.A.s who may, with ingenuity, have managed to escape all history of every sort. Our government and politics are peopled with such as these.

One of the most telling techniques taught by Neustadt and May is what they call "the Goldberg rule." The name comes from the head of a chain of grocery stores, Avram Goldberg, who told the authors, "When a manager comes to me, I don't ask him, 'What's the problem?' I say, 'Tell me the story.' That way I find out what the problem *really* is." Behind every issue is a story, the authors insist, and that story should begin with the earliest date that seems at all significant. An example given in the book concerns American debate in the 01980s about the rate of inflation. Studies that began after 01973 showed 10 to 12 percent inflation rates as "normal." Studies reaching back to 01953 showed those rates to be extraordinarily high. A two-century study would have shown that the United States had no inflation at all from the 01780s clear to the 01930s. In the light of the full story, "normal" is not 12 percent, but zero. High inflation was not the problem; any inflation was the problem.

In the 01970s my generation thought that solar water heaters on roofs were a good idea, and that cocaine was a fairly harmless drug. These opinions were policy in the Carter White House. We were blithely ignorant of American experience only a half century in the past, when solar water heaters routinely crashed through the newly rotted roofs of homes in Pasadena in the 01920s. And cocaine use was so virulent an epidemic in the 1910s that the government instituted a harsh and successful prohibition of the drug. Having failed to remember, we condemned ourselves to repeat that history on an even sorrier scale.

The past is both a comfort and a warning. It has to be both. If it is just a comfort, we become tranquilized and turn away from the future. If the past is just a warning, we may overlearn its lesson and seek a discontinuous break with the past, which then is bound to fail. Embracing the warning of the past along with its comfort is the essence of tragic optimism. With sufficiently unblinking hindsight, foresight may go well.

REFRAMING THE PROBLEMS

In 01996 a suddenly growing multibillion-dollar California foundation asked me and others to write a short paper on the question, "What are the most serious environmental problems confronting humankind at the beginning of the twenty-first century?" Figuring I would have nothing original to add to that list, I decided to write the piece from the perspective of the Clock of the Long Now. Looking from outside the present time gave a sideways rather than end-on view of the current environmental problems and invited rethinking them in terms of eventual practical solutions rather than only how great a threat they pose. I think the paper fits in at this point in the discussion, where the uses and advantages of long-view thinking are explored. The foundation (now the third-largest in America) is endowed with the wealth of David Packard, cofounder of Hewlett-Packard, the brilliantly successful electronics firm based in Palo Alto, California.

To the David and Lucile Packard Foundation:

My contribution may be to bend your question a little.

Environmental problems these days come in a pretty familiar litany of pretty familiar names. The World Population problem. Climate Change problem. Loss of Biodiversity. Ocean Fisheries. Freshwater Aquifers. North/South Economic Disparity. Rain Forests. Agricultural and Industrial Pollution. Identifying these issues and making them everyone's concern has been a major triumph of environmental science and activism in the late twentieth century.

I propose that the Packard Foundation could make a contribution beyond even the splendid effect of its funding by helping to rethink—reframe—the very structure of how environmental problems are stated. This is a common practice among inventive engineers such as the late Mr. Packard. When a design problem resists solution, reframe the problem in such a way that it invites solution.

An example of spontaneous reframing occurred in 01969, when the Apollo program began returning color photographs of the Earth from space. Everyone saw the photographs and saw that we occupied a planet that was beautiful, all one, very finite, and possibly fragile. The environmental movement took off from that moment—the first Earth Day was in 01970. That effect of the American space program was never intended or anticipated. Indeed, nearly all environmentalists in the sixties (except Jacques Cousteau) actively fought against the space program, saying that we had to solve Earth's problems before exploring space.

What might be some further helpful reframings?

(1) *Civilization's shortening attention span is mismatched with the pace of environmental problems.*

What with accelerating technology and the short-horizon perspective that goes with burgeoning market economies (next quarter) and the spread of democracy (next election), we have a situation where steady but gradual environmental degradation escapes our notice. The slow, inexorable pace of ecological and climatic cycles and lag times bear no relation to the hasty cycles and lag times of human attention, decision, and action. We can't slow down all of human behavior, and shouldn't, but we might slow down parts.

Now is the period in which people feel they live and act and have responsibility. For most of us *now* is about a week, sometimes a year. For some traditional tribes in the American northeast and Australia *now* is seven generations back and forward (175 years each direction). Just as the Earth photographs gave us a sense of *the big here*, we need things that give people a sense of *the long now*.

Candidate now-lengtheners might include: abiding charismatic artifacts; extreme longitudinal scientific studies; very large, slow, ambitious projects; human life extension (with delayable childbearing); some highly durable institutions; reward systems for slow responsible behavior; honoring patience and sometimes disdaining rush; widespread personal feeling for the span of history; planning practices that preserve options for the future.

In a sense, the task here is to make the world safe for hurry by slowing some parts way down.

(2) *Natural systems can be thought of pragmatically as "natural infrastructure."*

One area in which governments and other institutions seem comfortable thinking in the long term is the realm of infrastructure, even though there is no formal economics of infrastructure benefits and costs. (There should be and could be.) We feel good about investing huge amounts in transportation systems, utility grids, and buildings.

Infrastructure thinking is directly transferable to natural systems. Lucky for us, we don't have to build the atmosphere that sustains us, the soils, the aquifers, the wild fisheries, the forests, the rich biological complexity that keeps the whole thing resilient. All we have to do is defend these systems—from ourselves. It doesn't take much money. It doesn't even take much knowledge, though knowledge certainly helps.

A bracing way to think about this matter would be to seriously take on the project of terraforming Mars—making it comfortable for life. Then think about reterraforming Earth if we lose the natural systems that previously built themselves here. The fact is that humans are now so powerful that we are in effect terraforming Earth. Rather poorly so far. We can't undo our power; it will only increase. We can terraform more intelligently—with a light, slow hand, and with the joy and pride that goes with huge infrastructure projects. Current efforts by the Army Corps of Engineers to restore the Florida Everglades, for example, have this quality.

(3) *Technology can be good for the environment.*

My old biology teacher, Paul Ehrlich, has a formula declaring that environmental degradation is proportional to "population times affluence times technology." It now appears that the coming of information technology is reversing that formula, so that better technology and more affluence leads to less environmental harm—*if* that is one of the goals of the society.

"Doing more with less"—Buckminster Fuller's "ephemeralization"—is creating vastly more efficient industrial and agricultural processes, with proportionately less impact on natural systems. It is also moving ever more of human activity into an *infosphere* less harmfully entwined with the biosphere.

Given its roots, the Packard Foundation is particularly well suited to evaluate and foster what a Buddhist engineer might call *right technology*. It would be helpful to assemble a roster of existing environmentally benign technologies. Satellites for communication and remote sensing come to mind. So does Jim Lovelock's gas chromatograph (which detected atmospheric chlorofluorocarbons)—invented for Hewlett-Packard, as I recall.

The foundation might support activities such as Eric Drexler's Foresight Institute, which is aiming to shape nanotechnology (molecular engineering) toward cultural and environmental responsibility. It might support services on the Internet that distribute information and discussion about the environmental impacts of new and anticipated technologies and their interactions. Good effects should be investigated as well as ill effects.

(4) *Feedback is the primary tool for tuning systems, especially at the natural/artificial interface.*

German military officers are required to eat what their troops eat and after they eat. That single tradition assures that everyone's meals are excellent and timely, and it enhances unit morale and respect for the officers. The feedback cycle is local and immediate, not routed through bureaucratic specialists or levels of hierarchy.

In similar fashion, factories, farms, and cities that pollute rivers and water tables could be required to release their outflows upstream of their own water intake rather than downstream.

The much-lamented "tragedy of the commons" is a classic case of pathological feedback—where each individual player is rewarded rather than punished for wasting the common resource. In fact, healthy self-governing commons systems are frequent in the world and in history, as examined in Elinor Ostrum's *Governing the Commons*. The commons she dissects include communally held mountain meadows and forests in Switzerland, irrigation cooperatives in Japan and Spain, and jointly managed fisheries in Turkey, Sri Lanka, and Nova Scotia. The successful ones are maintained (and maintainable) neither by the state nor the market but by a local set of community feedbacks adroitly tuned to ensure the system's long-term health and prosperity. Ostrum detects eight design principles that keep a wide variety of commons self-balancing. They are: clear boundaries; locally appropriate rules; collective agreement; monitoring; graduated sanctions; conflict-resolution mechanisms; rights to organize; nested enterprises.

The Packard Foundation could encourage feedback analysis of environmental problems and help devise local-feedback solutions.

(5) *Environmental health requires peace, prosperity, and continuity.*

War, especially civil war, destroys the environment and displaces caring for the environment for generations. Widespread poverty destroys the environment and undermines all ability to think and act for the long term.

Environmental activists and peace activists are still catching on that they are natural partners, and both remain averse to business boosters who might aid prosperity. Peacekeeping soldiers are not in the mix at all. But for a culture and its environment to come into abiding equanimity you need all four—eco-activists, peace activists, marketeers, and honest cops—each of them with a light touch, comfort with collaboration, and eagerness to replace themselves with local talent. An example of productive joining of regional business and environmental goals is the Ecotrust project at Willapa Bay, Washington.

By its funding choices and guidelines Packard Foundation could foster "jointness" in world-saving endeavors. In support of the long now, it could promote people, ideas, and organizations that are in for the long haul.

The benefits of very long-term scientific studies are so obvious it is hard to understand why they are so rare.

Global warming, the dominant environmental issue of our time, might not be an issue at all but for a study measuring atmospheric carbon dioxide begun in 01958 in a Hawaiian hut by Charles Keeling and Roger Revelle. High on the slopes of Mauna Loa volcano, downwind of thousands of miles of Pacific Ocean, the instruments have shown a steady forty-year increase in the human-exacerbated greenhouse gas, CO_2, from 315 to 362 parts per million. Fateful numbers! Since they are largely the result of the aggregate metabolism of civilization, the trend will be an enormous task to reverse, but if it is not reversed, civilization faces a drastically different Earth over the coming century.

Maintained through four decades of budget worries and changes in scientific fashions, the Mauna Loa CO_2 records show the beginning effects of global warming (thus proof of it) as well as one of its major causes. You can see an annual cycle in the chart, with atmospheric carbon dioxide levels going down in the spring, when northern-hemisphere plants take up carbon for their growth spurt, and then rising again in the fall, when decay takes over. The *amplitude* of this cycle has increased some 20 percent in the forty years, indicating that Earth is "breathing deeper." The probable cause is a gradual overall increase in vegetation, fed by the higher CO_2 levels and perhaps stimulated by higher temperatures. Greenhouse indeed.

Enormous, inexorable power is in the long trends, but we cannot measure them or even notice them without doing extremely patient science. These days science is more often driven at commercial or even fashion velocity than at the deliberate pace of governance or the even slower pace of nature. As history accelerates, people become fast learners, and that's good, but it is also a problem. "Fast learners tend to track noisy signals too closely and to confuse themselves by making changes before the effects of previous actions are clear," says

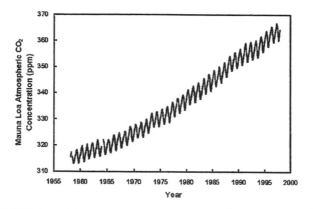

FIGURE 21.1 The recent increase in the greenhouse gas carbon dioxide is shown in this renowned graph from forty years of readings atop the Mauna Loa volcano in Hawaii. It is the prime evidence for humanity's involvement in causing global warming. The vertical dimension of the graph shows atmospheric carbon dioxide in parts per million. The rise from 01958 to 01998 is from 315 ppm to 362 ppm—an increase of 16 percent. Each year there is a seasonal oscillation, shown by the wiggles in the curve, which is driven by northern hemisphere plants taking up large quantities of CO_2 in their spring growth and releasing it in the decay of autumn.

decision analyst James March. Quiz shows and classroom teachers reward the quick answerer. This is not helpful in domains where the quick answer is the wrong answer.

A nine-year study in Africa concluded that burning new woody growth in open grassland could not prevent the woods from taking over. A forty year study of the same subject proved the opposite, that annual burning was an ideal way to keep the grasslands open. It takes more than a decade of fires to keep woody rootstocks from resprouting, that's all.

Nearly half of ecological field research spans only one year. The two longest animal studies are George Schaller's Serengeti lion research—twenty-seven years so far—and Jane Goodall's work on chimpanzees—thirty-six years. The longest time-lapse film (such as speeded-up flower opening) covers just a week. No one has yet studied the entire life span of a termite nest, which may extend to a hundred years or more, with several queens reigning in succession.

Really extended studies are highly uncommon. The world's longest and most fruitful agricultural research (later understood as also ecological) was begun in 01843 at the Rothamsted estate, near London. John Bennett Lawes and Joseph Henry Gilbert worked together there for fifty-seven years, producing one hundred fifty scientific papers and three hundred popular articles, establishing themselves as the founders of scientific agriculture. Their original studies continue to this day at the Rothamsted Experimental Station, more than a century and a half later. What began as a series of experiments on the nutrient requirements of crop plants soon answered those questions but raised a whole new set of questions about plant diversity, soil development, community ecology, and even evolution. The longer the data set grew the more valuable it became. Particularly crucial was the preserving of one hundred fifty years of soil and plant samples, which could later be examined with tools and lines of inquiry (such as pollutant analysis) unimagined by the founding scientists.

Rigorously collected old data keeps finding new uses. In 01790 the U.S. Constitution instituted the world's first regular population census of a nation to ensure that the population-based House of Representatives reflected accurate figures. This record later provided a precise profile of the growth, movement, and changing composition of the American people. In the 01990s the extensive marshlands of the south end of San Francisco Bay began being restored to their original teeming richness, thanks to an exquisitely detailed early map of the area by the 19th-century cartographer David Kerr. Sometimes natural systems can be mined for invaluable data sequences: polar ice cores showing the composition of the atmosphere for millennia past, tree-ring studies covering three thousand years in the American southwest and eight thousand years in Nepal, woodrat middens in Nevada preserving thirty-three thousand years of seed and pollen samples in columns of amberlike rat excrement.

One ecologist, Jim Brown, is trying to reverse the trend toward ever smaller and shorter field research projects by founding the study of *macroecology*, focusing on "phenomena at regional to global spatial scales and decadal to millennial temporal scales." In that perspective the otherwise diverging disciplines of ecology, biogeography, paleobiology, and macroevolution are forced to come together and make sense of each other. Effective macroecology relies not on short, clever experiments but on patient observation, correlation, and statistical analysis. Brown points out that this level of study is essential now that human impact is global in scale: "Ecologists have studied the effects of starfish, largemouth bass, sea otters, beavers, and other 'keystone' species, but they are strangely reluctant to study the most key species of all, their own." (An example: there were once ninety-nine species of land birds in Hawaii. The arrival of Polynesians removed fifty of them, the arrival of Europeans another seventeen, with nineteen more now in great danger of extinction, which leaves only thirteen of the original ninety-nine bird species intact in the company of humans.)

So in light of their great accumulative value, why *are* long-term scientific studies so rare? Well, (1) they're not about proving or disproving hypotheses, the coin of the scientific realm; (2) they don't generate quick papers, the coin of a scientific career; (3) they bear no relation to scientific fashion, where the excitement is; (4) they're not subject to money-making patent or copyright; (5) the few that exist usually die when their primary researcher dies; (6) they're extremely difficult to maintain funding for; and (7) ever-growing archives are an expensive hassle to service and keep accessible ("We can't stop the future to take care of the past!").

One morning at the Santa Fe Institute in New Mexico I was having breakfast with Stanford geneticist Marcus Feldman. He said that universities are dying to do long-term research, but they can't count on reliable sources of funding over time. He then described a project of his: studying the new and very troubling gender imbalance of babies in China—119 boys born for every 100 girls! Due to the customary preference for male children and the government ban on large families, mothers are getting illegal ultrasound tests and aborting female fetuses. Feldman's team had found an unusual Chinese town where the ratio of boys to girls was still even, apparently because the local custom was for newlyweds to live with the wife's parents instead of the husband's. Feldman selected an otherwise comparable town with the usual 119:100 ratio and set in motion a long-term demographic study comparing what happens over time in the two towns. "We need at least twenty years of follow-through to see what the real effects are," he fretted.

At that moment we were joined by Robert Galvin, the longtime head of Motorola Corporation. I asked Galvin how his company's major bet on China was going. He waxed enthusiastic about their grand initiative, and he was intrigued to hear about Feldman's research in China: "Those children you're studying are going to be our employees in twenty years. We plan to be in China for hundreds of years. I think you should get in touch with Motorola's foundation."

If a Long Now Library gets established, one useful role for it might be to broker ambitious longitudinal science studies with deep-pocketed—or steady-pocketed—funding sources. It might also guarantee long-term oversight and archival backup for the studies. When they are abandoned by their original researchers, it could try to find new keepers of the work, or at least preserve the accumulated material for later review or revival. Such a Library could foster cross-pollination among the long-term projects: correlating data and spreading the word on new tools, new uses for old data, and newly evolved best practices.

Science and art are always inspiring each other. Maybe some works of slow art could shame science into durational ambition. Paul Saffo likes hiding Easter eggs: "a brilliant ceramic sculpture hidden within rocklike concrete which slowly weathers away to reveal a 'Hi there!' from another era. Something wonderful buried in a flood plain where the river snakes back and forth and in time will carve a bank into the treasure. One could develop the genre to where lots of people do it for a while, making the world a very slow, very amazing Advent calendar."

THE
LONG VIEW

Caught up as we are these years in the whirligig of time, with our attention-deficit disorder and our technological obsession with the ever tinier and ever faster, how do we keep up with its pace and at the same time perceive outside it? Supposing that occasionally taking the long, contemplative view is indeed a good thing, where do you stand to get one? One authority, Dr. Gregory Cajete, a Tewa Indian from the American Southwest, has this advice:

> The elders remind us of the importance of the long view when they say, "pin peyeh obe"—look to the mountain. They use this phrase to remind us that we need to look at things as if we are looking out from the top of a mountain, seeing things in the much broader perspective of the generations that are yet to come. They remind us that in dealing with the landscape, we must think in terms of a ten-thousand-, twenty-thousand-, or thirty-thousand-year relationship.

One day in the late sixties humanity found itself atop a new peak, the Moon, viewing Earth and Earth's history from an altitude of 240,000 miles. It turned out that the astronomer Fred Hoyle was right in 01947 when he forecast, "Once a photograph of the Earth, taken from outside, is available . . . a new idea as powerful as any in history will be let loose." In those photos from the Moon we saw fractured humanity's home as itself unfractured and whole, and we began to see what the Tewas saw from local mountains: a relationship measured in millennia. This insight will keep being renewed as space tourism develops in the coming decades.

The environmentalist René Dubos also was right: "We are becoming planetized probably almost as fast as the planet is becoming humanized." Our global influence and our global perspective are almost keeping pace with each other (which is fortunate—it could have been otherwise). Once we acknowledge our new responsibility for the health of the planet, the large view and the long view become one. The Big Here and the Long Now merge as the Long Here, which is no longer just occupied but managed by what might be called the Long Us. The Chinese have a term for it: *da wo*, or "big me."

There are other vivid points of perspective on the Long Us. Religions collect and protect many. Kevin Kelly reported on a visit to the Greek peninsula of monasteries known as Mount Athos.

It is very remote, very stable, very conservative, very much still working. There's maybe two dozen different monasteries, all breathtaking in their architecture and settings. I'd have to say it was the most timeless place I've been to. You couldn't tell what century it was at the moment. The monasteries house and feed you (men only) if you show up—that's their rule. Those meals, with the monks in full hooded garb, by candlelight, and in complete silence, were stories in themselves. I don't know what drew them to that place, but they were running the best time machine I know of.

Paul Saffo took his deepest time out as a youth:

I spent two summers living among a community of Maya in the highlands of Chiapas and two summers at Palenque working on the decipherment of archaeoastronomical hieroglyphs. I found all this a perfect counterpoint to what I was doing in the winters, studying history of science and technology policy. At Copan, a site with some of the most impressive inscribed stone slabs, there was one with a long count date of more than four billion years. The Maya clearly knew how to celebrate the dance of deep time.

Reader, what was the occasion of your longest view?

Communal time has been the norm over most of human history, says anthropologist Steve Barnett, and "individual time" only came in with the European Enlightenment. "I knew a guy in south India," Barnett remarked, "who looked across a field and told me that his ancestors had been farming there, in that place, for three thousand years. He knew his ancestors back eight generations." Long here, long us.

The slow stuff is the serious stuff, but it is invisible to us quick learners. Our senses and our thinking habits are tuned to what is sudden, and oblivious to anything gradual. Between the near-impossible win of a lottery and the certain win of earning compound interest, we choose the lottery because it is sudden. The difference between fast news and slow nonnews is what makes gambling addictive. Winning is an *event* that we notice and base our behavior on, while the relentless losing, losing, losing is a nonevent, inspiring no particular behavior, and so we miss the real event, which is that to gamble is to lose.

What happens fast is illusion, what happens slow is reality. The job of the long view is to penetrate illusion.

Edward Gibbon's *Decline and Fall of the Roman Empire* excelled in this kind of perception; it's right there in the title. For Gibbon, writes Robert D. Kaplan,

> the more gradual and hidden the change, the more important it turned out to be. . . . The real changes were the insidious transformations: Rome moving from democracy to the trappings of democracy to military rule; Milan in Italy and Nicomedia in Asia Minor functioning as capital cities decades before the formal division of the empire into western and eastern halves, and almost two centuries before Rome ceased to be an imperial capital; the fact that the first fifteen "Christian" bishops of Jerusalem were circumcised Jews subscribing to a not yet formalized religion.

Is Gibbon's view only possible from the distance he had, of fifteen centuries? How can we see the insidious transformations of our own day? Slow science is part of it, applied history is part of it, and every year there are more sophisticated tools of macroscopic vision. One video going the rounds of the conferences shows the accelerating growth of human population on a world map; the sudden overwhelm in the last seconds makes audiences gasp in shock.

I know field biologists who can look at a hillside and "see" the advance of scrub growth over failing meadow; look at a wide valley and see the river lashing like a snake within its floodplain, the meander loops progressing downstream and flicking off oxbow slues to either side; look at a terminal moraine like Cape Cod and see the glacial ice advance and then withdraw over the landscape to a one-hundred-thousand-year beat. That kind of ability is made of knowledge absorbed until it becomes perception.

Prolonged observation and deep perspective is something we assign to institutions. They aggregate human effort and are expected to handle such durational tasks as noticing slow, important change. Even some corporations, commerce-hasty as they are, do it: responsible lumber companies farming their woods on seventy-year cycles, insurance companies with their century-scale actuarial tables. But governments and universities are the main institutions we charge with caring for the long view: Governments owing to their level of responsibility, universities because their duties are to intellectual heritage and the ever-new generations passing through.

The great advisor on institutional management, Rosabeth Moss Kanter, notes that

> people care about their place in history when their own past is valued. . . . People take the long view when they perceive leaders as trustworthy. . . . [They] take the long view when they believe the rules of the game are fair. They believe they will share equitably in the returns. . . . [They] take the long view when they have a deep understanding of system dynamics. They see the connections between actions in one place and consequences in another. They can therefore appreciate the need for indirect long-term investments (whether research and development, infrastructure repairs, or education).

Kanter concludes, "People take the long view when they feel a commitment to those who come after them. . . . They care about posterity—their children and other people's children—and therefore see the need for actions to benefit the distant future."

The long view looks right through death.

GENERATIONS

When you've been through enough winters you finally come to know and truly believe, in the dark of the year, that spring will come—just when you're no longer sure that you'll be around for it. Unlike the young, for whom each season is a world, the old can savor the passing of the seasons, actually feel them move through, charged with poignancy.

The great problem with the future is that we die there. This is why it is so hard to take the future personally, especially the longer future, because that world is suffused with our absence. Its very life emphasizes our helpless death. The power granted to humans by foresight is enormous, but so is the cost. We can plan only in the bitter knowledge of personal extinction. Shakespeare said it: "Thought's the slave of life, and life time's fool; And time, that takes survey of all the world, Must have a stop. O! I could prophesy, But that the earthy and cold hand of death lies on my tongue."

Time will not have a stop; it won't even slow down. That may explain why people have been speeding up, as if by cramming more and more life into each passing hour they can personally enact Zeno's Paradox, always never more than halfway to death. Our technology does offer us various forms of life compression, from jet planes and cell phones to stimulants and multitasking tools. If *chronos* is the problem, day-grabbing *kairos* looks like the solution. In some degree it may well be.

Yet each person's portion of *chronos*—our lifespan—in fact has been increasing dramatically. Global human life expectancy at birth was about ten years throughout most of human history. With safer childbirth and improvements in medicine, a newborn's life expectancy reached thirty-four in 01900, forty-six in 01954, and sixty-four in 01998. Even without a medical breakthrough it is expected to be seventy-two by 02020. In one lifetime life expectancy has increased 50 percent. The number of Americans over age sixty-five has gone from 4 percent in 01900 to 12 percent in the late 01990s to perhaps 20 percent by the 02020s.

In one century elders have gone from being rare and honored to common and powerful. The most dominant lobby in Washington, D.C., is The American Association of Retired People. Never in history have so many generations been alive at the same time. Living long enough to know your great-grandchildren has become the norm, even with delayed childbearing. Among the things that the new elders are doing with their power—and their accumulated wealth—is directing ever more sophisticated research toward life extension. I have heard biotech scientists seriously ask one another, off the record, "What if we cure death?" Whether or not effective immortality actually comes, its prospect is now in sight, and that itself begins to change things.

The best science fiction on life extension is Bruce Sterling's *Holy Fire* (01996). In its deliberately stable world dominated by the "medical-industrial complex," one character explains, "When you live a really long time, it changes everything. The whole structure of the world, politics, money, religion, culture, everything that used to be human. All those changes are your responsibility, they benefited you, they happened because of you. You have to work hard so that the polity can manage."

If long life leads to greater responsibility, because you hang around long enough to suffer the consequences of your short-sighted actions, then immortality logically leads to infinite responsibility. Esther Dyson once mused, "I personally like the certainty of death. It is amazingly relaxing to realize that one can't do everything. If I knew I were going to live forever, I would feel obliged to fix all my imperfections. I would have to learn many more languages; I would worry about my teeth not holding out; I would have to make amends for all the mistakes I have made." Kevin Kelly responded, "That's why those religions that believe in eternal virtual life are so hung up on perfecting and repenting."

Organic farmers have a bumper sticker: "Live like you'll die tomorrow. Farm like you'll live forever." When effective immortality kicks in, we all become organic farmers in one form or another, and that's not so bad. But it will do strange things to the interaction among generations, which is already pretty strange . . .

"Once an angry man dragged his father along the ground through his own orchard," wrote Gertrude Stein. ". . .'Stop!' cried the groaning old man at last. 'Stop! I did not drag my father beyond this tree.'"

When Brian Eno approached the father of Anthea Norman-Taylor for permission to marry her, he was told, "What you have to ask yourself is, 'Would I wish this woman to be the grandmother of my grandchildren?'"

How about great-great-great-great grandchildren, every one of them with orchard-dragging potential? Proliferating elders may have their own bumper sticker: "Never trust anyone under 50." The very old will have experienced enough past to believe in the reality of consequences, while the young will not have been wrong about enough future yet to doubt their own puerile notions. It's the old dialogue, but with a new balance of power when the old outnumber the young.

The bonds between immediate family generations may loosen, while the overall bond among humanity's generations becomes stronger, simply because so many generations are sharing the same world—having direct experience of the Long Us. Esther Dyson again: "As I get older, my 'age group' widens on both sides. When I was small I felt a certain kinship with five-year-olds, but six-year-olds were of another generation, and four-year-olds were little punks. When I was a teenager, my range extended a year or two in either direction. Now, in my mid-forties, my 'generation' includes people who grew up or even fought in the Second World War."

That's one scenario, inclusive and probably rather conservative. Another scenario, far more divisive, could be built around the continuing acceleration of technology: In this world only the young may be able to keep up, and thus they become the ones wielding overwhelming power. They may not be happy to have their world cluttered with out-of-it oldsters. Society could fragment into mutually hostile age cohorts, with the younger ones having no interest whatever in the old orchards, and lighting out for Mars.

A third scenario, perhaps the ideal outcome, would be for the bonds between generations to grow stronger, with an ever-growing bias toward the young. Danny Hillis recalls, "What my grandfather did was create options. He worked hard to allow my father to have a better education than he did, and in turn my father did the same." The American revolutionary John Adams wrote to his wife in 01780:

I must study politics and war that my sons may have liberty to study mathematics and philosophy. My sons ought to study mathematics and philosophy, geography, natural history, naval architecture, navigation, commerce, and agriculture, in order to give their children a right to study painting, poetry, music, architecture, statuary, tapestry, and porcelain.

Speaking of porcelain, in *The Travels of Marco Polo* is an account of a generation-spanning practice in thirteenth-century coastal China:

In a city called Tinju, they make bowls of porcelain, large and small, of incomparable beauty. They are made nowhere else except in this city, and from here are exported all over the world. . . . These dishes are made of a crumbly earth or clay which is dug as though from a mine and then stacked in huge mounds and left for thirty or forty years exposed to wind, rain, and sun. By this time the earth is so refined that dishes made of it are of an azure tint with a very brilliant sheen. You must understand that when a man makes a mound of this earth he does so for his children; the time of maturing is so long that he cannot hope to draw any profit from it himself or put it to use, but the son who succeeds him will reap the fruit.

The preserving of options for future generations can be progressive, as in Adams's case, or conservative, keeping a valuable environment, tool, or material intact and ripening. Those things were done for us. The debt we cannot repay our ancestors we pay our descendants.

SUSTAINED ENDEAVOR

Earthquakes, war, murder, the burning of libraries . . . *bad things happen fast*. Reforestation, the growth of a child, the maturing of an adult, building a library . . . *good things happen slow*.

That's not universally true, of course, but it's true enough to wonder if there is a structural explanation. I think there are several. One is simple: Construction usually requires sequential elements of assembly, whereas destruction can be done all at once. Another explanation is embodied in all the folk wisdom about the carelessness of haste, such as the fable of the giddy Hare and the steady Tortoise, "A hasty man drinks tea with a fork" (Chinese proverb), and "Haste makes waste, and waste makes want, and want makes strife between the goodman and his wife" (seventeenth-century English proverb).

Early researchers in artificial intelligence came across yet another explanation, a phenomenon they called *hill climbing*. Imagine a mountain range of opportunities, where the higher you get the greater the advantage. Hasty opportunists will never get past the foothills because they only pay attention to the slope of the ground under their feet, climb quickly to the immediate hilltop, and get stuck there. Patient opportunists take the longer view to the distant peaks, and toil through many ups and downs on the long trek to the heights.

There are two ways to make systems fault-tolerant: One is to make them small, so that correction is local and quick; the other is to make them slow, so that correction has time to permeate the system. When you proceed too rapidly with something mistakes cascade, whereas when you proceed slowly the mistakes instruct. Gradual, incremental projects engage the full power of learning and discovery, and they are able to back out of problems. Gradually emergent processes get steadily better over time, while quickly imposed processes often get worse over time.

The astonishing sophistication of ancient poems such as *The Iliad*, *The Odyssey*, and *Beowulf* long has baffled scholars. How could

Homer be such a genius? Recent study of illiterate bards in our own day shows that they are always partially improvising for every performance, which solves the problem. The genius of "Homer" was the accumulated ideas of generations of bardic improvisation. *The Iliad* is so effective because it is so highly evolved. Likewise, science truly took off in the seventeenth century when the Royal Society introduced the idea of the scientific "letter" (now "paper"), which encouraged a torrent of small, incremental additions to scientific knowledge.

Except for open-ended endeavors like science, the tremendously powerful lever of time has seldom been employed. The pyramids of Egypt and Central America took only fifty years to build. Some of the great cathedrals of Europe indeed were built over centuries, but that was due to funding problems rather than patience. Humanity's heroic goals generally have been sought through quick, spectacular action ("We will land a man on the Moon in this decade") instead of a sustained accumulation of smaller, distributed efforts that might have overwhelming effect over time. The kinds of goals that can be reached quickly are rather limited, and work on them displaces attention and effort that might be spent on worthier, longer-term goals.

Danny Hillis points out, "There are problems that are impossible if you think about them in two-years terms—which everyone does—but they're easy if you think in fifty-year terms." This category of problems includes nearly all the great ones of our time: The growing disparities between haves and have nots, widespread hunger, dwindling freshwater resources, ethnic conflict, global organized crime, loss of biodiversity, and so on. Such problems were slow to arrive, and they can only be solved at their own pace. It is the job of slow-but-steady governance and culture to set the goals of solving these problems and to maintain the constancy and patience required to see them through (that is not our current model of governance).

Restorative goals such as these are the most important, but they do have a negative cast. Could their accomplishment be aided by also engaging some positive goals that operate at the same pace? Colonizing Mars has this quality. Building a 10,000-Year Clock/Library might. Assembling a universal virtual-reality world on the Net feels like an achievable great work. Success in mapping the human genome should encourage the related ambition of inventorying all the species on Earth and mapping their genomes. Filling in all the gaps and blanks in the total human family tree would be a vivid experience of the Long Us.

These are first-thought blurts. We have not yet seriously asked ourselves what we might do with fifty years or five hundred years of sustained endeavor. What comes to your mind, thinking in that scale?

It is not easy achieving such things. This is part of the attraction, that the task is impossible-seeming and bracingly hard. Herman Melville wailed during the writing of *Moby Dick*, "Oh, Time, Strength, Cash, and Patience!" The rewards of immersion in a project, a story, reaching well beyond the span of one's own life, however, can be enormous. This is some of what keeps people working gladly in long-lived institutions such as universities and religions; it would be the main attraction of very-long-term science studies.

Environmental projects, owing to the extended lag times involved and perhaps the aesthetic rewards along the way, excel at inspiring long-term ambition. I know of two North American environmental projects with thousand-year time frames: Ecotrust, which is setting about building a nature-sustaining economy throughout the coastal temperate rain forest from mid-California to northern Alaska; and The Wildlands Project, which aims to restore enough wild land, surrounded by partially wild "buffer zones" and connected by wildlife migration corridors, for native animal and plant populations to survive indefinitely amid human dominance of the continent. Instead of saving endangered species individually and temporarily, the idea is to take the time to save them all permanently.

The learning theorist Seymour Papert tells of a group of friends eating lobsters at a Boston fish house. The question came up, "Can anyone eat lobster without making a mess?" Papert reports, "A brain surgeon at the table did it. It took him two hours—a completely eaten lobster with a perfect absence of mess. He took the time appropriate to the job, which he knew about. It wasn't his skill. It was his patience."

Two hours was the difference between impossible and easy. For what tasks would two hundred years make that kind of difference?

It was a professor of religion, James P. Carse of New York University, who came up with the idea of "the infinite game." His jewel of a book, *Finite and Infinite Games* (01986), begins, "A finite game is played for the purpose of winning, an infinite game for the purpose of continuing the game."

Football, elections, and much of business are finite games: win/lose. Family, gardening, and spiritual practice are infinite games: Losing is meaningless. Finite games, Carse points out, require fixed rules so that the winner and loser are determined fairly, but infinite games thrive on occasional changes in the rules—agreed to by the players—so that the game constantly improves. Finite players seek to control the future; infinite players arrange things so the future keeps providing surprises. Death-defying finite players seek immortality through their famous victories; infinite players "offer their death as a way of continuing the play—they do not play for their own life, they live for their own play."

FINITE GAME

The purpose is to win
Improves through fittest surviving
Winners exclude losers
Winner takes all
Aims are identical
Relative simplicity
Rules fixed in advance
Rules resemble debating contests
Compete for mature markets
Short-term decisive contests

INFINITE GAME

The purpose is to improve the game
Improves through game evolving
Winners teach losers better plays
Winning widely shared
Aims are diverse
Relative complexity
Rules changed by agreement
Rules resemble grammar of original utterances
Grow new markets
Long term

The above is from a management book applying Carse's theory, *Mastering the Infinite Game*, by Charles Hampden-Turner and Fons Trompenaars, 01997.

The distinction recalls the ancient Greek differentiation between opportunity-grabbing *kairos* and all-inclusive *chronos*. Greek tragedy probed exactly such a contrast, says management consultant Charles Hampden-Turner: "Medea does not really kill her own children. She pretends to, on stage, because she is so furious with Jason, their father. And the audience surely got the message that quarreling parents kill the lives of their children by degrees. All Greek tragedies had the same structure. The finite game of feuding heroes and heroines proves itself so deadly that it kills the infinite game of parental nurture, royal succession, wise governance, etc."

Infinite games are corrupted by inappropriate finite play. Governance (infinite) is disabled when factional combat (finite) becomes the whole point instead of providing helpful debate and alternation of power. Cultures (infinite) perish when one culture seeks to eradicate another. Nature (infinite) is dangerously disrupted when commercial competition (finite) lays waste to natural cycles. Finite games flourish *within* infinite games, but they must not displace them, or all the games are over.

Finite games focus on how they end, while infinite games focus on how they continue. Freeman Dyson, a Princeton mathematician who grew up and was educated in England, once told me that the habit of long-term thinking "survives all over England. It's one reason the country has been so amazingly well cleaned up after the

Industrial Revolution. When I was a boy, I went to London, and my clothes were filthy at the end of the day. The city was covered with soot and grime, and the rivers were very polluted; it's all been cleaned up in the past fifty years. You can always improve things as long as you're prepared to wait."

Though long based in America, Dyson is still a member of Trinity College, Cambridge, which, he notes, "has been a fantastic producer of great science for four hundred years and continues to be so." I protested that scientific revolutions turn over whole previous constructs of the universe; how could that kind of activity take place in those same buildings over four centuries? "It goes naturally together," Dyson replied. "You need the space of continuity to have the confidence not to be afraid of revolutions."

That bears repeating. You need the space of continuity to have the confidence not to be afraid of revolutions. You can always improve things as long as you're prepared to wait.

Infinite play yields strange dividends. Corfu, once the usual goat-bitten barren Greek island, was encouraged by the mercantile empire of Venice to plant olive trees for a period of four hundred years and became the most fertile and beautiful of all the Greek isles. Another island, Visingsö, in the Swedish lake Vättern, has a gorgeous mature oak forest whose origin came to light in 01980 when the Swedish Navy received a letter from the Forestry Department reporting that the requested ship lumber was now ready. It turned out that in 01829 the Swedish Parliament, recognizing that it takes one hundred fifty years for oaks to mature and anticipating that there would be a shortage of timbers for its navy in the 01990s, ordered twenty thousand trees to be planted and protected for the navy. Lone opposition came from the Bishop of Strängnäs, who said he had no doubt that there would still be war in the late twentieth century but that ships might be built of other materials by then. In finite-game terms he was right, and worthy of remembrance. For the infinite game of healthy forests it is fortunate that he was ignored by wrongheaded, long-thinking Parliament.

Infinite play converges on other infinite play. Veterans returning from the First World War were treated badly in America. In 01944 it was those aging veterans, then in politically conservative

American Legion posts, who pushed through the GI Bill for returning World War II veterans, providing them with college tuition and low-cost home mortgages; it was not a Roosevelt New Deal program at all. The GI Bill's cost of $14.5 billion was paid back eightfold in taxes in the next twenty years, it jump-started the boom years of the 01950s, it built the world's largest middle class, and it set the nation decades ahead as the world moved into a knowledge economy. America's greatest infrastructural investment ever was made as a gesture of gratitude and justice rather than of profound forethought. A move in one infinite game—generational responsibility—paid off in another infinite game—growing prosperity. Perhaps James Carse is right to end his book with the words, "There is but one infinite game."

Maturity is largely a combination of hard-earned savvy, the habit of thinking ahead, and the patience to see long-term projects through. If *kairos* is for the young, *chronos* is the domain of mature individuals and societies. Their embrace of duration yields wisdom, described by the scientist Jonas Salk as "the capability of making retrospective judgments prospectively." Wisdom decides forward as if back. Rather than make detailed, brittle plans for the future, wisdom puts its effort into expanding general, adaptive options. A fertile Corfu has more options than a barren one; veterans with a college education had more options than those without. An Earth with an intact ozone layer has more options than one without.

Preserving and increasing options is a major component of a self-saving world. Making it a habit would be part of the answer to the question, How do we make long-term thinking automatic and common instead of difficult and rare? Time-inclusive thinking began when the first farmers planted their seeds instead of eating them (it must have seemed a risky investment). The story of civilization is the story of ever-new forms of thinking ahead and the results of those forms. How the story will play out we have no way of knowing. The product of even the most imaginative and prudent forethought is not certainty but surprise. This is the reward for infinite-game generosity. Surprise plus memory equals learning. Endless surprise, diligent memory, endless learning.

While I was completing this book, the poet Gary Snyder sent me an epigram that had come to him:

THE CLOCK OF THE LONG NOW

> *This present moment*
> *That lives on to become*
> *Long ago.*

I felt it was *The Clock of the Long Now* that responded to him:

> *This present moment*
> *Used to be*
> *The unimaginable future.*

APPENDIX:
ENGAGING
CLOCK/LIBRARY

What the Clock/Library project offers its contributors is the possibility of being instructed and amused—maybe even inspired—for the rest of their lives. At minimum it is the work of many decades. In that time no end of ideas should be tried for their fit with each other and with the Long Now mission of expanding humanity's sense of time and responsibility.

The quality of the project—the depth of its originality and the reach of its impact—will depend entirely on how it is nourished in terms of ideas, alliances and money. If you would like to participate, you can reach The Long Now Foundation via its web site at: www.longnow.org or through its headquarters in the San Francisco Presidio:

The Long Now Foundation
P.O. Box 29462
The Presidio
San Francisco, CA 94129
Phone: (415) 561–6582
Fax: (415) 561–6297

Long Now is a 501 (c) 3 nonprofit education foundation; donations are tax-deductible (tax ID number: 68–0384748). On the web site cash gifts can be made by credit card, and there is a list of additional current needs such as particular equipment and specialized services. Volunteer help also is welcome.

The other coin of this project is ideas. What do you think would make a 10,000-Year Clock more effective, and a 10,000-Year Library more useful? Send in your thoughts.

The project gains most of its resource leverage through alliances with other organizations. So far these have included the National Park Service, Disney, Global Business Network, and The Getty Center. If your organization would like to engage some aspect of the Clock or the Library, please get in touch.

Long Now's executive director, Alexander Rose, can be contacted by E-mail at zander@longnow.org, and I am at sb@gbn.org.

—Stewart Brand
January 01999

The first working prototype of the 10,000-year Clock as it appeared on December 31, 01999, to bong in the new millennium. It already has a monumental quality. When you stand in front of it, the lower rim of the Clock face is eye level, and you look up to the year date on the right and the central starfield with its overlay of curving lines showing where Earth's rotational axis currently points. (Photo by Rolfe Horn)

AFTERWORD:
JANUARY 02000

One tick of the Clock later, what can be reported?

In twelve months Long Now built the first working Clock, found a desert mountain to build the big one in, and took first steps toward the Library. It happened because a surprising number of remarkable people delighted in taking the long-term seriously and personally, and used Long Now as a convenient avenue for that.

The Library effort got a major boost when Michael Keller, head of libraries at Stanford University, joined the Long Now board. Then Charlie Butcher, founder of a foundation called Lazy 8, put up $140,000 to fund two Library projects in 02000. At Stanford in June there would be a conference on "The 10,000-year Library." A group of luminaries would work up from the question, "What does civilization need?" rather than down from the question, "What do libraries need?"

The second Library project is a "Rosetta Disk"—a profusion of Norsam micro-micro-etched nickel disks capable of retaining their information in microscope-readable form for thousands of years. On the disk will be the most-translated text in the world—probably "Genesis" from the Old Testament—in hundreds of ancient and contemporary languages, along with other Creation stories and additional aids to language scholars of the future. The project leader is a land artist and anthropologist named Jim Mason.

Construction of the Prototype Clock swarmed ahead on schedule. At one point Danny Hillis remarked, "This is a clock that Harrison or Babbage could have built, if they'd had FedEx." Indeed the eighteenth-century creator of the Longitude clock and the nine-

teenth-century builder of the first mechanical computer would recognize the crafting details of the eight-foot-high device, though
the bit adder rings replacing gear ratios would be new (and fascinating) to them. They would smile at the familiar brass and marvel
at the Monel ("the alloy that stainless steel would like to be"). They
would marvel even more at Federal Express's ability to bring precisely machined—and redesigned and remachined—parts from distant fabricators overnight to Chris Rand's high-tech machine shop
in a decaying shipyard on the waterfront of Sausalito, California.

There Danny Hillis's clock design was refined and improved by
designer (and project manager) Alexander Rose, mechanical engineer Liz Woods, and clockmaker David Munroe. The result is a
work of great technical achievement and considerable beauty.

Despite its easily understood display of only Sun, Moon, day
length, starfield, and year date, by some ways of reckoning it is the
most complex clock ever built. For instance, it transparently carries
out: the complicated formula for Leap Years over centuries; the
ever-shifting Equation of Time relationship between solar time
and local time; plus the way that relationship keeps changing because of the 25,764-year Precession of the Equinoxes; and the effect of that Precession on the locally visible starfield. Because the
Prototype Clock is meant to travel, the day-length display is ad-
justable for a variety of Northern Hemisphere latitudes. The Clock
is calming to watch because the rotating pendulum takes a full (and
exact) minute to complete one cycle—one tick of the escapement.
Once every hour energy is released from a power column to drive
the five adder rings—clickety click, clickety click—to perform the
digital calculations that, when appropriate, advance the display
rings.

On New Year's Eve 01999 a small group gathered to watch how
the just-assembled Clock would handle the millennial turn. At
midnight the governor balls on one power column spun up, the
adder rings rotated, then TWO hexagonal Geneva wheels rotated
and up on the Clock face "019" gave way to "020" and "99" to
"00." A preliminary cheer was shushed as a hidden Zen Buddhist
gong inside the Clock base declared:

Bonnnnnnngggggggggg
and
Bonnnnnnnggggggggggggggg · · ·

Such a fine sustain in the ring of it, keeping attention held. Then the Clockwrights were grinning and shaking hands with each other, in relief and respect and joy. In only a thousand years we could come back and hear it bong three times. Meanwhile the Clock went back to the shop for fifty fixes and improvements—that's the job of a prototype—that Hillis wanted to make before the Clock performed publicly at a large conference in February. In June 02000 the Clock would travel on loan to London's renowned Science Museum to climax a new permanent exhibit on the history of technology. If $2-million funding could be found, a larger—twelve-foot-high—finished Clock would later take up perpetual residence. The exhibit's director, Andrew Nahum, said the museum could promise good care of the Clock for centuries.

But what about millennia? Long Now's search for a desert mountain site for an underground Clock of serious longevity was expected to take years or decades.

In fall 01998 Roger Kennedy, who had been director of the National Park Service from 01993 to 01997, led a group of Long Now board members and friends to the Service's newest and least-known park, Great Basin National Park in eastern Nevada. Small in size for a western park, its main attractions are a popular cave and the highest mountains in Nevada, the Snake Range. Those mountains have a steep west-facing escarpment, and Mount Washington in particular (elevation 11,636 feet) had spectacular 2,000-foot cliffs of highly durable, brilliant white limestone. According to maps there was a large piece of private property next to park land amid those shining cliffs.

The geology was impressive but the main attractions were the people of White Pine County and its main town, Ely (population 5,000). Everyone the Long Now group met was devoted to the spare, remote landscape and its history, yet they were also intrigued by the Clock idea. Park Superintendent Becky Mills was immediately supportive. The real estate agent who showed the group around, Dave Tilford, turned out to be a man of considerable vision—and skill. Six months of intense negotiation with the defunct mining firm that owned the property finally yielded a workable price of $140,000 for 180 acres.

It was a fair offer, but the terms were cash, now. In just a couple of weeks the Mitchell Kapor Foundation, the family of Jay Walker (founder of Priceline.com), and Sun Microsystems co-founder Bill Joy came up with the money. The deal was done. Long Now had a foothold on a plausible, exciting 10,000-year Clock site.

The most spectacular feature of Mount Washington is its forest of ancient bristlecone pine trees. They're big, tough as stone, each one gnawed into a unique sculpture by the high-altitude winters. The young ones are a thousand years old. The oldest in the Snake Range (and the world) dates back 4,900 years—then the dense tree rings blur into old heartwood. Rings in some dead bristlecone trunks on the mountain go back 10,000 years.

When showing visitors around the mountain Dave Tilford advises, "We've found it's a good idea for people to just go sit by themselves for five minutes here. Find a bristlecone you like and hang out with it. Forget the present. Think about the past and the future." I've seen him do that half a dozen times. People always take longer than five minutes. When you see them again, they're quiet.

The gateway town for Great Basin National Park is tiny Baker, Nevada, just off U.S. 50—known as "America's Loneliest Highway." In August 01999 Alexander Rose and I were there to announce Long Now's land purchase. Most of the twenty or so people in the room had a copy of this book. Their questions were friendly but sharp (as yours might be).

"When do you think the Clock will be finished?"

"If it's finished in my lifetime," said 28-year-old Alexander, "we're doing it wrong."

"Stewart, how has working on this project affected you personally?"

That one stopped me. The questioner was JoAnne Garrett, a well-respected environmentalist with a bright eye. "It's made getting old fun," I reported, discovering the fact as I spoke.

After that meeting and a larger one in Ely an old rancher, tough as bristlecone himself, Mr. Baker, came up to me and Alexander. "You know," he said, "I came to the meeting yesterday with just one question about this thing you're trying to do: 'Why?' The question I have now is: 'Why not?' "

His wry grin brightened winters to come.

Top: Mt. Washington in eastern Nevada has impressive 2,000-foot-high Cambrian limestone cliffs that may provide a 10,000-year home for a large and durable Clock of the Long Now. If the project is funded and continues to be welcomed by White Pine County, the Clock would be built inside the cliffs, so this view in a hundred or a thousand years would look the same as it does now.

Below: More than most mountains, Mt. Washington is a time machine because of the extensive forest of bristlecone pine trees near its peak. This photo by Stephen Johnson shows the branches of a dead tree weathered by millennia of winters and the growth of a young bristlecone—only centuries old—behind it. Behind that is a hundred miles of Nevada desert. Hikers on the mountain can't help experiencing the long view, in every sense.

NOTES

All quotations in the text which are not cited in these notes come from personal communications with the author, usually in the form of E-mail.

CHAPTER 2, *KAIROS* AND *CHRONOS*

:08 **12.6 billion new humans will be having lives. . . .** This number has apparently never been derived before, since demographers have been asked to focus on numbers of people alive, rather than individuals having lives. Chris Ertel, a demographer at Global Business Network, made the calculation. He started with the 1998 United Nations population estimate for 2100 C.E.—10.4 billion. He then added the number of people likely to have died during the 21st century—8.3 billion (calculated by accounting for the effects of UN-estimated death rates through the century, in five-year increments). Then he subtracted the UN-estimated population of the world on January 1, 2000 C.E.—6.1 billion. Thus the number of new human lives in the period 2000–2100 C.E. is: 10.4 billion + 8.3 billion – 6.1 billion = 12.6 billion.

:09 **"We are the first generation that influences global climate . . ."** O. Orheim, "The Norwegian Glacier Centre" (publicity pamphlet, 1992), quoted in Graham May, *The Future Is Ours* (Westport, CT: Praeger, 1996), p.79.

:09 **"We are changing Earth more rapidly than we are understanding it."** Peter Vitousek et al., "Human Domination of Earth's Ecosystems," *Science* (25 July 1997), p. 498.

:09 **"*kairos* (opportunity or the propitious moment) and *chronos* (eternal or ongoing time)"** Patricia Fortini Brown, *Venice and Antiquity* (New Haven, CT: Yale Univ., 1996), p. 6.

CHAPTER 3, MOORE'S WALL

:15 **"We are moving from a world in which the big eat the small . . ."** Quoted in column by Thomas Friedman, *New York Times* (13 November 1996), p. A19.

:16 **"What people mean by the word *technology* . . ."** The remarks by Alan Kay and Danny Hillis are frequently made by them in speeches.

:16 **"The world did not double or treble its movement between 1800 and 1900 . . ."** Quoted by Arthur Schlesinger Jr., "Has Democracy a Future?" *Foreign Affairs* (September 1997), p. 5.

:16 **"Five-year plan???"** "Managing on (Internet) Time," *Wired* (June 1998), p. 86.

:16 **"If Moore's Law is true, over time is time more or less valuable?"** Luyen Chou, president and CEO of Learn Technologies Interactive in New York.

:17 **"continuous discontinuous change"** Regis McKenna, *Real Time* (Cambridge, MA: Harvard Business School, 1997)

:17 **"Some people say that they feel the future is slipping away . . ."** Danny Hillis, "The Millennium Clock," *Wired Scenarios* (1995), p. 48.

CHAPTER 4, THE SINGULARITY

:20 **"At this singularity the laws of science and our ability to predict the future would break down."** Stephen Hawking, *The Illustrated A Brief History of Time* (New York: Bantam, 1988, 1996), p. 114.

:20 **". . . a place where extrapolation breaks down and new models must be applied"** Vernor Vinge, *Across Realtime* (Riverdale, NY: Baen, 1991), p. 402.

CHAPTER 5, RUSH

:25 **"SPREAD OF TECHNOLOGY GIVES RISE TO A CULTURE OF IMMEDIACY"** *Christian Science Monitor* (5 March 1998), p. 5.

:25 **"Imagine a world in which time seems to vanish . . ."** Regis McKenna, *Real Time* (Cambridge, MA: Harvard Business School, 1997)

:25 **the rise of gambling. . .** The statistics on gambling are assembled from: Robert Goodman, *The Luck Business* (New York: Free Press, 1995); *Science* (23 June 1998), p. 4; Evan I. Schwartz, "Wanna Bet," *Wired* (October 1995), p. 134.

CHAPTER 6, THE LONG NOW

:29 **"When I pronounce the word Future . . ."** Wislawa Szymborska, "The Three Oddest Words," *New York Times Magazine* (1 December 1996), p. 49.

:29 **"If one is mentally out of breath all the time from dealing with the present . . ."** Elise Boulding, "The Dynamics of Imaging Futures," *World Future Society Bulletin* (September 1978), p. 7.

CHAPTER 7, THE ORDER OF CIVILIZATION

:34 **R. V. O'Neill and C. S. Holling.** R. V. O'Neill, D. L. DeAngelis, J. B. Wade, and T. F. H. Allen, *A Hierarchical Concept of Ecosystems* (Princeton, NJ: Princeton Univ., 1986). C. S. Holling, "What Barriers? What Bridges?" *Barriers & Bridges to the Renewal of Ecosystems and Institutions* (New York: Columbia Univ., 1995).

:35 **"Consider a coniferous forest.** C. S. Holling, title above, p. 23.

:35 **"The destiny of our species is shaped by the imperatives of survival on six distinct time scales."** Freeman Dyson, *From Eros to Gaia* (New York: Pantheon, 1992), p 341.

:36 **"Every form of civilization is a wise equilibrium . . ."** Eugen Rosenstock-Huessy, *Out of Revolution* (Norwich, VT: Argo, 1938, 1969), p. 563.

:38 **"The social sector."** Peter Drucker, the cultural historian and business scholar, is responsible for this coinage.

:38 **The world's first empire, the Akkadian in the Tigris-Euphrates valley, lasted only a hundred years. . .** Richard A. Kerr, "Sea-Floor Dust Shows Drought Felled Akkadian Empire," *Science* (16 January 1998), p. 325.

CHAPTER 8, OLD-TIME RELIGION

:42 **"They are not following time, but sustaining it . . ."** Michel Serres, *The Natural Contract* (Ann Arbor, MI: The Univ. of Michigan, 1992), p. 47.

:43 **"This age, known as Very Beautiful, Very Beautiful, lasted 400 trillion oceans of years . . ."** Quoted in Joseph Campbell, *The Masks of God, Vol. 3: Oriental Mythology* (Arkana, 1991).

:43 **"The breathing in and out of the universe by Brahma every four thousand million years is not an image of the future . . ."** Elise Boulding, "The Dynamics of Imaging Futures," *World Future Society Bulletin* (September 1978), p. 6.

:43 **"... their revolt against concrete, historical time ..."** Mircea Eliade, *The Myth of the Eternal Return* (Princeton, N.J.: Princeton Univ., 1949, 1954), p. ix.

:43 **The Pharaoh's job was to maintain eternal order...** These observations about Egypt's timelessness come from Daniel Boorstin, *The Creators* (New York: Random, 1992), pp. 156–8.

CHAPTER 9, CLOCK/LIBRARY

:48 **"In some sense, we've run out of our story ..."** Quoted in Po Bronson, "The Long Now," *Wired* (May 1998), p. 118.

:48 **"If you're going to do something that's meant to be interesting for ten millennia ..."** Danny Hillis interview with Richard Kadrey, *HotWired* (5 December 1995).

:51 **"To me the Clock and the Library capture two different aspects of time."** Danny Hillis in same *HotWired* interview as above.

:53 **Historian Daniel Boorstin reports that the Inner Shrine at Ise...** Daniel Boorstin, *The Creators* (New York: Random House, 1992), p. 140.

CHAPTER 10, BEN IS BIG

:56 **The monument is named for its biggest bell, Big Ben.** Most of the information in this chapter comes from John Darwin, *The Triumphs of Big Ben* (London: Hale, 1986).

CHAPTER 11, THE WORLD'S SLOWEST COMPUTER

:68 **"Within the human cortex crouches an impulse to build something huge."** Kathryn Gabriel, *The Roads to Center Place* (Boulder, CO: Johnson, 1991), p. 1.

CHAPTER 12, BURNING LIBRARIES

:72 **"Oh, Septimus!—can you bear it?"** Tom Stoppard, *Arcadia* (London: Faber & Faber, 1996), p. 38.

:72 **The legendary stature of the Library of Alexandria is justifiable.** My sources on the contents, working, and burnings of the Library of Alexandria included: Michael H. Harris, *History of Libraries in the Western World* (Metuchen, NJ: Scarecrow, 1995), pp. 42–51; Luciano Canfora, *The Vanished Library* (Berkeley, CA: UC California, 1987); L.D. Reynolds, N.G. Wilson, *Scribes and Scholars* (Oxford, England: Clarendon, 1991), pp. 6–8; Daniel Boorstin, *The Creators* (New York: Random, 1992), p. 47.

:73 **"If their content is in accordance with the book of Allah . . ."** Luciano Canfora, *The Vanished Library* (Berkeley: UC California, 1987), pp. 98–99.

:74 **A different reason for burning books was originally invented by China's first great emperor.** My best source on Shih huang-ti's book burning was one of Time-Life's excellent *Lost Civilizations* series, *China's Buried Kingdoms* (Alexandria, VA: Time-Life, 1993). The dinner debate is reported on p. 98.

:74 **The same impulse inspired Hitler's book-burning ceremonies of May 1933.** William L. Shirer, *The Rise and Fall of the Third Reich* (New York: Simon & Schuster, 1960), pp. 230, 241.

:74 **The American Revolution of 1776, by contrast, was highly conservative.** The American historian Samuel Eliot Morison made this argument persuasively. His 1975 speech, "The Conservative American Revolution," is in a Morison anthology: Emily Morison Beck, *Sailor Historian* (Boston: Houghton-Mifflin, 1977), pp. 234–253.

:75 **The Spanish Conquest missionaries burned the codices of the Mayans.** My main sources on the Mayan losses were: Michael D. Coe, *The Maya* (New York: Thames & Hudson, 1966, 1987), p. 161; *The Magnificent Maya* (Alexandria, VA: Time-Life, 1993), pp. 15–17, 29.

:75 **For three days Serb forces targeted Sarajevo's multicultural National and University Library with a bombardment. . .** Len A. Costa, "The Libraries: Another Kind of War Victim," *New York Times* (13 January 1998), p. A15; Diane Asseo Griliches, *Library* (Univ. of New Mexico, 1996), p. 114.

CHAPTER 13, DEAD HAND

:78 **"Christians preferred to be buried in the sanctified ground of the graveyard, creating layer upon layer of putrefying parishioners . . ."** Alice Outwater, *Water* (New York: Basic, 1996), p. 135.

:79 **"I do not see how future ages are to stagger onward under all this dead weight . . ."** Nathaniel Hawthorne, quoted in Jeannette Greenfield, *The Return of Cultural Treasures* (Cambridge Univ., 1989), p. 310.

:79 **"America, you have it better than our old continent . . ."** Johann Wolfgang Goethe, quoted in *Irresistible Decay* (Los Angeles: Getty, 1997), p. 55.

:79 **"Layer upon layer, past times preserve themselves in the city . . ."** Lewis Mumford, *The Culture of Cities* (New York: Harcourt, Brace, 1938)

:79 **Each new U.S. president leaves behind more papers to be preserved than all the previous presidents combined.** Personal communication from David Lowenthal, author of *Possessed by the Past* and *The Past Is a Foreign Country*.

CHAPTER 14, ENDING THE DIGITAL DARK AGE

:82 **"Digital information is forever."** Andrew Grove, quoted in Regis McKenna, *Real Time* (Cambridge, MA: Harvard, 1996).

:82 **"It is only slightly facetious to say that digital information lasts forever—or five years, whichever comes first."** Jeff Rothenberg, "Ensuring the Longevity of Digital Documents," *Scientific American* (January 1995), p. 42.

:85 **. . . computer professionals call these monsters *legacy systems.*** There is a good discussion of legacy systems by Ellen Ullman, *Close to the Machine* (San Francisco: City Lights, 1997), pp. 116–8.

:86 **In 1090 C.E. the Chinese genius Su Sung built a monumental water-driven mechanical clock. . .** David Landes, *Revolution in Time* (Cambridge, MA: Harvard, 1983), pp. 29–36; Daniel Boorstin, *The Discoverers* (New York: Penguin, 1983), pp. 59–61.

:87 **How much information exists in the world today. . . ?** Michael Lesk, "How Much Information Is There in the World?" (BellCore, 1997), www.gii.getty.edu/timeandbits/ksg.html.

CHAPTER 15, TEN-THOUSAND-YEAR LIBRARY

:94 **"What has been done, thought, written, or spoken is not culture . . ."** Gary Taylor, *Cultural Selection* (New York: Basic, 1996), p. 6.

:96 **. . . as Patrick Ball has done with archives in Salvador, Guatemala, and South Africa, working from his base in Washington, D.C.** Todd Lapin, *Wired* (January 1998), p. 105.

:96 **One such man, Ken McVay, undermined all the online Holocaust-denying discussion groups. . .** McVay's Nizkor Project is at www.nizkor.netizen.org.

:97 **"It is an unusually moving thing to initiate a message which will not be read until long after one's death."** Richard Slaughter, "Why We Should Care for Future Generations Now," *Futures* (October 1994), p. 1083.

:97 **The Biological and Environmental Specimen Time Capsule 2001. . .** Yasuko Kamizumi, "In Deep Freeze, a Little Air and DNA for the Future," *New York Times* (21 February 1998), p. A15.

:99 **. . . the convicted Dr. Sheppard was innocent back in 01954. . .** Fox Butterfield, "DNA Test Absolves Sam Sheppard of Murder, Lawyer Says," *New York Times* (5 March 1998), p. A14.

1:01 **Only one copy of Lucretius made it through the Dark Age. . .** L. D. Reynolds, N. G. Wilson, *Scribes and Scholars* (Oxford:

Clarendon, 1991), p. 101; Gary Taylor, *Cultural Selection* (New York: Basic, 1996), p. 77.

1:01 **"The death of a language . . ."** George Steiner, *Errata* (New Haven, CT: Yale Univ., 1998), p. 114.

1:02 **The scientist James Lovelock has proposed compiling a start-up manual for civilization. . .** James Lovelock, "A Book for All Seasons," *Science* (8 May 1998), pp. 832–3.

1:03 **The largest and heaviest book in the world is inscribed on 14,300 large stone tablets. . .** Arlette Kouwenhoven, "Largest, Heaviest Book," *Archaeology* (May 1996).

CHAPTER 16, TRAGIC OPTIMISM

1:07 **"Much was decided before you were born."** This is one of artist Jenny Holzer's "truisms."

1:08 **Gregg Easterbrook has written a whole fat book of environmental good news. . .** Gregg Easterbrook, *A Moment on the Earth* (New York: Viking, 1995).

1:08 ***The Wooing of Earth*** René Dubos, *The Wooing of Earth* (New York: Scribners, 1980).

1:09 ***The Idea of Decline in Western History*,** Arthur Herman, *The Idea of Decline in Western History* (New York: Free Press, 1997).

1:09 **"Is the glass half-empty, or half-full?"** Bill Cosby, quoted in the *New Haven Register* (14 May 1995), p. A2.

CHAPTER 17, FUTURISMO

1:12 **"He refused to pay her price," one history reports.** "Sibyl," Encyclopedia Britannica Online (www.eb.com).

1:13 **"There are two sets of futures," Desmond Bernal wrote in 01929. . .** quoted in Freeman Dyson, *Infinite in All Directions* (New York: Harper, 1989), p. 291.

1:14 **"Economic forecasting makes predictions by extrapolating curves of growth from the past into the future."** Freeman Dyson, Ibid., p. 180.

1:15 **"Economic forecasting misses the real future because it has too short a range."** Freeman Dyson, Ibid., p. 180.

CHAPTER 18, USES OF THE FUTURE

1:19 **The shift from the de Klerk to the Mandela government was greatly eased by what became known as the Mont Fleur Scenarios.** Available from The Institute for Social Development,

University of the Western Cape, Private Bag X17, 7535 Bellville, South Africa.

1:19 **Our mental framework has what philosopher Derek Parfit describes as a "bias toward the future."** Derek Parfit, *Reason and Persons* (Oxford Univ., 1984), p. 160.

1:22 **"The proper role of government in capitalistic societies in an era of man-made brain power industries is to represent the interest of the future to the present."** Lester Thurow, *The Future of Capitalism* (New York: Penguin, 1997), p. 16.

1:22 **A major point of reference for thinkers about the future is Robert Axelrod's 01984 book,** *The Evolution of Cooperation.* Robert Axelrod, *The Evolution of Cooperation* (New York: Basic, 1984).

CHAPTER 19, USES OF THE PAST

1:26 **"Most Europeans live in towns and villages which existed in the lifetime of St. Thomas Aquinas . . ."** George Holmes, *Introduction to The Oxford Illustrated History of Medieval Europe* (Oxford Univ., 1988), p. v.

1:27 **At Laetoli, Tanzania, archaeologist Mary Leakey discovered an impossibly ancient trail of hominid footprints. . .** Mary Leakey, quoted by Neville Agnew and Martha Demors, "The Footprints at Laetoli," *Conservation* (Spring 1995), p. 16.

1:27 **"I love everything that's old . . ."** Oliver Goldsmith, *She Stoops to Conquer,* Act I.

1:28 **"This storm is what we call progress."** Walter Benjamin, "Theses on the Philosophy of History" (1940), *Illuminations* (New York: Schocken, 1968), p. 257.

1:28 **"Those who cannot remember the past are condemned to repeat it."** George Santayana, *The Life of Reason; Vol. 1: Reason in Common Sense* (1905) (New York: Prometheus, 1998).

1:28 **Describing Winston Churchill, Isaiah Berlin wrote. . .** Isaiah Berlin, "Winston Churchill in 1940," (1949) *The Proper Study of Mankind* (New York: Farrar Straus, 1998), p. 608.

1:29 **"General knowledge of history is less and less characteristic of American decision-makers and their aides."** Richard Neustadt and Ernest May, *Thinking in Time* (New York: Free Press, 1988), p. 245.

CHAPTER 20, REFRAMING THE PROBLEMS

1:35 **Elinor Ostrum's *Governing the Commons*.** Elinor Ostrum, *Governing the Commons* (Cambridge Univ., 1990).

CHAPTER 21, SLOW SCIENCE

1:38 The *amplitude* of this cycle has increased some 20 percent in the forty years, indicating that Earth is "breathing deeper." Bill McKibben, "A Special Moment in History," *The Atlantic Monthly* (May 1998), p. 68.

1:38 "Fast learners tend to track noisy signals too closely and to confuse themselves by making changes before the effects of previous actions are clear." James G. March, *A Primer on Decision Making* (New York: Free Press, 1994), p. 245.

1:39 A forty-year study of the same subject proved the opposite. . . M. J. Swift, et al,. "Long-term Experiments in Africa," in R. A. Leigh and A. E. Johnston, *Long-term Experiments in Agricultural and Ecological Sciences* (Wallingford, UK: CAB International, 1994), p. 244.

1:40 The world's longest and most fruitful agricultural research. . . Leigh and Johnston, Ibid., pp. 9–39; 117–39.

1:40 . . . the study of *macroecology*, focusing on "phenomena at regional to global spatial scales and decadal to millennial temporal scales." James H. Brown, *Macroecology* (Chicago, IL: Univ. of Chicago, 1995), pp. 19, 205, 209.

CHAPTER 22, THE LONG VIEW

1:44 "The elders remind us of the importance of the long view . . ." in Norbert S. Hill, Jr., *Words of Power* (Fulcrum, 1994).

1:44 "We are becoming planetized probably almost as fast as the planet is becoming humanized." René Dubos, *The Wooing of Earth* (New York: Scribners, 1980), p. 70.

1:46 ". . . the more gradual and hidden the change, the more important it turned out to be." Robert D. Kaplan, "And Now for the News," *The Atlantic Monthly* (March 1997), p. 18.

1:47 "People care about their place in history when their own past is valued." Rosabeth Moss Kanter, *On the Frontiers of Management* (Cambridge, MA: Harvard, 1997), pp. 281–2, 284.

CHAPTER 23, GENERATIONS

1:50 Yet, each person's portion of *chronos*—our lifespan—in fact has been increasing dramatically. The statistics in this paragraph come from: Marshall N. Carter and William G. Shipman, "The Coming Global Pension Crisis," *Foreign Affairs* (November 1996), p. 98; Laura Carstenson, *Stanford Alumni Magazine* (March 1998), p. 48; and John W. Rowe, Robert L. Kahn, *Successful Aging* (New York: Pantheon, 1998), pp. 3, 6.

1:51 **"When you live a really long time, it changes everything."**
Bruce Sterling, *Holy Fire* (New York: Bantam, 1996), p. 36.

1:52 **"Once an angry man dragged his father along the ground through his own orchard . . ."** Gertrude Stein, *The Making of Americans* (1925; Dalkey Archive Press, 1998), opening line.

1:53 **"I must study politics and war that my sons may have liberty to study mathematics and philosophy."** John Adams, letter to Abigail Adams, May 12, 1780.

1:53 **"In a city called Tinju, they make bowls of porcelain, large and small, of incomparable beauty."** Marco Polo, *The Travels of Marco Polo* (ca. 1300; New York: Penguin, 1958), p. 238.

CHAPTER 24, SUSTAINED ENDEAVOR

1:56 **"A hasty man drinks tea with a fork."** Both sayings from Stuart and Doris Flexner, *Wise Words & Wives' Tales* (New York: Avon, 1993), p. 79.

1:57 **The genius of "Homer" was the accumulated ideas of generations of bardic improvisation.** Daniel Boorstin, *The Creators* (New York: Random, 1992), p. 29.

1:57 **The pyramids of Egypt and Central America took only fifty years to build.** Flora Simmons Clancy, *Pyramids* (Washington, DC: Smithsonian, 1994), p. 106.

1:58 **I know of two North American environmental projects with thousand-year time frames.** Ecotrust, 1200 NW Naito Pkwy., Ste. 470 Portland, Oregon; www.ecotrust.org. The Wildlands Project, 1955 W. Grant Rd., Ste. 148A, Tucson, Arizona; www.twp. org.

CHAPTER 25, THE INFINITE GAME

1:60 **The infinite game. . .** James P. Carse, *Finite and Infinite Games* (New York: Ballantine, 1986). Charles Hampden-Turner and Fons Trompenaars, *Mastering the Infinite Game* (Oxford, UK: Capstone, 1997).

1:62 **"You can always improve things as long as you're prepared to wait."** Interview by Stewart Brand, "Freeman Dyson's Brain," *Wired* (February 1998), p. 173.

1:62 **Another island, Visingsö, in the Swedish lake Vättern, has a gorgeous mature oak forest whose origin came to light in 01980. . .** The story was reported in 1994 in the Swedish newspaper *Svenska Dagbladet* and on the Swedish TV news show, *Rapport*.

1:63 **The GI Bill's cost of $14.5 billion was paid back eightfold in taxes in the next twenty years. . .** Michael J. Bennett, *When Dreams Came True* (Washington, DC: Brassey's, 1996). Milton Greenberg, *The GI Bill* (New York: Lickle, 1997).

RECOMMENDED BIBLIOGRAPHY

Many books informed *The Clock of the Long Now*. These are the ones I loved.

The Myth of the Eternal Return, Mircea Eliade (Princeton, NJ: Princeton Univ., 1949, 1954)
Humanity's love of timelessness.

Guns, Germs, and Steel, Jared Diamond (New York: Norton, 1997)
Biogeography made agriculture, made civilization.

Time Frame, Charles Boyle, ed., 24 volumes, (Richmond, VA: Time-Life, 1990)
The 10,000 year story.

The Complete Pyramids, Mark Lehner (London: Thames & Hudson, 1997)
The most durable signal through time.

Lost Civilizations, Dale M. Brown, ed., 23 volumes (Richmond, VA: Time-Life, 1993)
Greatness falls, time after time.

The Civilization of the Middle Ages, Norman F. Canter (New York: HarperCollins, 1963, 1968, 1974, 1993)
Surviving a dark age.

The Discoverers, Daniel Boorstin (New York: Penguin, 1983)
How ideas drove history.

The Creators, Daniel Boorstin (New York: Penguin, 1992)
Art, invention, and science through history.

The Vanished Library, Luciano Canfora (New York: Random House, 1992)
The Library of Alexandria.

How the Irish Saved Civilization, Thomas Cahill (New York: Doubleday, 1995)
Heroics of continuity.

Revolutions in Time, David Landes (Cambridge, MA: Harvard, 1983)
How clocks synchronized civilization.

The Triumphs of Big Ben, John Darwin (London: Hale, 1986)
A clock becomes a nation.

The Pattern on the Stone, W. Daniel Hillis (New York: Basic Books, 1998)
The primordial computer.

The Idea of Decline in Western Civilization, Arthur Herman (New York: Free Press, 1997)
Doom is a bad idea.

Possessed by the Past, David Lowenthal (New York: Free Press, 1996)
History as burden.

Thinking in Time, Richard Neustadt and Ernest May (New York: Free Press, 1988)
Applying history well.

The Wooing of Earth, René Dubos (New York: Scribners, 1980)
Tending nature.

Built to Last, James C. Collins, Jerry I. Porras (New York: Harper, 1994)
Organizing for longevity.

The Evolution of Cooperation, Robert Axelrod (New York: Basic, 1984)
Time builds trust.

Governing the Commons, Elinor Ostrum (New York: Cambridge Univ., 1990)
Untragic common sense.

Holy Fire, Bruce Sterling (New York: Bantam, 1996)
Long life changes life.

Infinite in All Directions, Freeman Dyson (New York: Harper, 1989)
  Sophisticated hope.

The Art of the Long View, Peter Schwartz (New York: Doubleday, 1991)
  Scenarios engage and free the future.

The Age of Spiritual Machines, Ray Kurzweil (New York: Viking, 1999)
  Beyond Moore's Law.

Deep Time, Gregory Benford (New York: Avon, 1999)
  Signalling through millennia.

Future Survey, Michael Marien, ed. (Monthly newsletter, $79/year from: World Future Society, 7910 Woodmont Ave., Suite 450, Bethesda, MD 20814, USA)
  All the good new books and articles about the future.

Encyclopedia Britannica Online (www.eb.com)
  Quick, deep online research.

Finite and Infinite Games, James P. Carse (New York: Ballantine, 1986)
  In the infinite game, to play is to win.

ACKNOWLEDGMENTS

Thanks, first of all, to John Brockman, literary agent, who instigated *The Clock of the Long Now* and the *Master Minds* series it joins. John is one of the great intellectual enzymes of our time. (An enzyme is a biological catalyst—an adroit enabler of otherwise impossible things.)

More than most books, this one came from conversation. The main source was discussion among the members of the board of The Long Now Foundation, most of it via thousands of messages online, a few of them quoted in this volume. Danny Hillis, Peter Schwartz, Brian Eno, Doug Carlston, Kevin Kelly, Paul Saffo, Mitch Kapor, Esther Dyson, executive director Alexander Rose and new board member Roger Kennedy *are* the Clock/Library project. It is no accident that nearly all of them are also members of Global Business Network (GBN), whose eleven-year development of "the art of the long view" stimulated and framed much of what the Long Now is about. GBN also is a conversation, for which I am especially grateful to its co-founders, Peter Schwartz, Jay Ogilvy, Napier Collyns and Lawrence Wilkinson. The conversation that made this book is neither exclusive nor over. I hope you'll join it. We have yet to make a world in which long-term thinking is automatic and common.

Reading a book in manuscript is an onerous task, and commenting on it to the author is a delicate matter. Those who did both include Danny Hillis, Kevin Kelly, Peter Schwartz, John Brockman, Brian Eno, Kees Van der Heijden, and Ryan Phelan. An early line-edit came from James Donnelly, and a later one from Michael Wilde. The primary, and very able, editor of the book is John Donatich at

Basic Books, New York, with editor Toby Mundy at Orion Books in London, a highly diligent and effective back-up. The beautiful cover was made by Nick Castle and Takumi in London.

Assistance of all kinds for the book came from: Jaron Lanier, Christina Gerber, Noah Johnson, Bill Joson, Paul Hawken, Tomi Pierce, Mattias Sôderhielm, Isabella Kirkland, Charles Hampden-Turner, Chris Ertel, and countless others, some of whom will glance at this list and quite rightly wince that they're not named—my oversight.

You may have noticed that male authors always reserve their ultimate gratitude (and apology) for their wives. There are good reasons for that, as my wife, Ryan Phelan, keeps learning. A particularly deep bow, and blush, to her.

INDEX

24143222R00131

Made in the USA
San Bernardino, CA
13 September 2015